Problem solving in physics is not simply a test of understanding the subject, but is an integral part of learning it. In this book, the basic ideas and methods of quantum mechanics are illustrated by means of a carefully chosen set of problems, complete with detailed, step-by-step solutions.

After a preliminary chapter on orders of magnitude, a variety of topics are covered, including the postulates of quantum mechanics, Schrödinger's equation, angular momentum, the hydrogen atom, the harmonic oscillator, spin, time-independent and time-dependent perturbation theory, the variational method, identical particles, multielectron atoms, transitions and scattering. Most of the chapters start with a summary of the relevant theory, outlining the required background for a given group of problems. Considerable emphasis is placed on examples from atomic, solid-state and nuclear physics, particularly in the latter part of the book as the student's familiarity with the concepts and techniques increases.

Throughout, the physical interpretation or application of the results is highlighted, thereby providing useful insights into a wide range of systems and phenomena. This approach will make the book invaluable to anyone taking an undergraduate course in quantum mechanics.

Problems in quantum mechanics
with solutions

Problems in quantum mechanics

with solutions

G. L. SQUIRES

Emeritus Lecturer in Physics at the University of Cambridge and Fellow of Trinity College, Cambridge

PUBLISHED BY THE PRESS SYNDICATE OF THE UNIVERSITY OF CAMBRIDGE
The Pitt Building, Trumpington Street, Cambridge, United Kingdom

CAMBRIDGE UNIVERSITY PRESS
The Edinburgh Building, Cambridge CB2 2RU, UK
40 West 20th Street, New York, NY 10011–4211, USA
477 Williamstown Road, Port Melbourne, VIC 3207, Australia
Ruiz de Alarcón 13, 28014 Madrid, Spain
Dock House, The Waterfront, Cape Town 8001, South Africa

http:/www.cambridge.org

First published 1995
Reprinted 1995, 1996, 2002

Typeset in Linotron Times 10/12.5 pt

A catalogue record for this book is available from the British Library

Library of Congress Cataloguing in Publication Data
Squires, G. L. (Gordon Leslie)
Problems in quantum mechanics with solutions / G. L. Squires.
p. cm.
Includes bibliographical references and index.
ISBN 0-521-37245-3. – ISBN 0-521-37850-8 (pbk.)
1. Quantum theory – Problems, exercises, etc. I. Title.
QC174. 15. S66 1995
530. 1′2 – dc20 93-43931 CIP

ISBN 0 521 37245 3 hardback
ISBN 0 521 37850 8 paperback

Transferred to digital printing 2003

KT

Contents

Acknowledgements

I started this book in 1969. I took sabbatical leave in 1970, at the Hebrew University in Jerusalem, intending to finish it there. However I was side-tracked by meeting the woman whom I subsequently married. So, like P. G. Wodehouse, I dedicate this book to my wife, but for whom it would have been written twenty three years ago.

I wish to thank Dr K. F. Riley, and Messrs S. R. Johnson, S. Patel, and B. E. Rafferty who read parts of the manuscript and made several useful comments on it. I am particularly grateful to Dr M. E. Cates and Messrs D. M. Freye, F. M. Grosche, R. K. W. Haselwimmer, R. J. F. Hughes, and C. S. Reynolds, who between them worked through the problems and made valuable comments and suggestions. Finally, I wish to acknowledge my indebtedness to all the undergraduates, mainly from Trinity College, whom I have supervised in Quantum Mechanics in the last thirty years. In the words of the proverb 'I have learnt much from my teachers, but more from my pupils.'

G. L. Squires

Preface for the reader

The problems in this book are intended to cover the topics in an average second- and third-year undergraduate course in Quantum Mechanics. After a preliminary chapter on orders of magnitude, there are eight chapters on topics arranged in a fairly conventional order. The tenth and final chapter contains a selection of miscellaneous problems on the topics of the previous chapters. I have separated them from the others on the grounds of their being somewhat longer and perhaps more difficult. But you should not be deterred from trying them on that account.

The important thing for all the problems is that you *do attempt them*. If you attempt a problem, and think about it, but cannot solve it, and *then* look up the solution, you will get much more benefit than if you jump to the solution as soon as you have read the problem. If you can solve a problem, you are still advised to look at the solution, which might contain a quicker or neater method than the one you have used. (If yours is quicker or neater I shall be pleased to hear from you.) I have also included some comments at the ends of some of the solutions, which you may find useful. They relate, either to the algebraic technique, or, more commonly, to a physical interpretation or application of the result.

At the beginnings of Chapters 2 to 9, I have included sections entitled Summary of theory, and you should read the summary before trying the problems in the chapter. The summary has a two-fold object. One is to introduce the notation, and the other is to inform you what you need to know before you attempt the problems. The results are quoted without proofs, which it is assumed you will obtain in your lecture course.

The equations are numbered independently in each solution and summary. A single equation number refers to the equation within the current solution or summary. An equation in another solution is referred to by a triple number, e.g. (5.7.3) is equation 3 in the solution to Problem 5.7. Reference from outside to an equation in a summary is made by a double number, so (4.8) is equation 8 in the summary for Chapter 4.

1
Numerical values

Values of physical constants

speed of light	c	$= 2.998 \times 10^{8}\ \mathrm{m\,s^{-1}}$
permittivity of vacuum	$\varepsilon_0 = 1/\mu_0 c^2$	$= 8.854 \times 10^{-12}\ \mathrm{F\,m^{-1}}$
Planck constant	h	$= 6.626 \times 10^{-34}\ \mathrm{J\,s}$
	$\hbar = h/2\pi$	$= 1.055 \times 10^{-34}\ \mathrm{J\,s}$
elementary charge	e	$= 1.602 \times 10^{-19}\ \mathrm{C}$
Boltzmann constant	k_B	$= 1.381 \times 10^{-23}\ \mathrm{J\,K^{-1}}$
Avogadro constant	N_A	$= 6.022 \times 10^{23}\ \mathrm{mol^{-1}}$
mass of electron	m_e	$= 9.109 \times 10^{-31}\ \mathrm{kg}$
mass of proton	m_p	$= 1.673 \times 10^{-27}\ \mathrm{kg}$
mass of neutron	m_n	$= 1.675 \times 10^{-27}\ \mathrm{kg}$
atomic mass unit	$m_u = 10^{-3}/N_A$	$= 1.661 \times 10^{-27}\ \mathrm{kg}$
Bohr radius	$a_0 = 4\pi\varepsilon_0\hbar^2/e^2 m_e$	$= 5.292 \times 10^{-11}\ \mathrm{m}$
Rydberg constant	$R_\infty = \hbar/4\pi c m_e a_0^2$	$= 1.097 \times 10^{7}\ \mathrm{m^{-1}}$
fine structure constant	$\alpha = e^2/4\pi\varepsilon_0 c\hbar$	$= 7.297 \times 10^{-3}$
Bohr magneton	$\mu_B = e\hbar/2m_e$	$= 9.274 \times 10^{-24}\ \mathrm{J\,T^{-1}}$
nuclear magneton	$\mu_N = e\hbar/2m_p$	$= 5.051 \times 10^{-27}\ \mathrm{J\,T^{-1}}$

The values of the physical constants given above are sufficiently precise for the calculations in the present book. In fact, these constants are known with a fractional error of 10^{-6} or less, apart from the Boltzmann constant where the fractional error is about 10^{-5}. A list of the values of the fundamental physical constants which are the best fit to the results of a variety of precision measurements has been prepared by Cohen and Taylor (1986).

Problems

The values of physical constants are given on p. 1. The answers to Problems 1.1 to 1.7 should be given to 3 significant digits.

1.1 The ionisation energy of the hydrogen atom in its ground state is $E_{\text{ion}} = 13.60$ eV. Calculate the frequency, wavelength, and wave number of the electromagnetic radiation that will just ionise the atom.

1.2 Atomic clocks are so stable the second is now *defined* as the duration of 9 192 631 770 periods of oscillation of the radiation corresponding to the transition between two closely spaced energy levels in the caesium-133 atom. Calculate the energy difference between the two levels in eV.

1.3 A He–Ne laser emits radiation with wavelength $\lambda = 633$ nm. How many photons are emitted per second by a laser with a power of 1 mW?

1.4 In the presence of a nucleus, the energy of a γ-ray photon can be converted into an electron–positron pair. Calculate the minimum energy of the photon in MeV for this process to occur. What is the frequency corresponding to this energy?
[The mass of the positron is equal to that of the electron.]

1.5 If a dc potential V is applied across two layers of superconducting material separated by a thin insulating barrier, an oscillating current of paired electrons passes between them by a tunnelling process. The frequency ν of the oscillation is given by $h\nu = 2eV$. Calculate the value of ν when a potential of 1 V is applied across the two superconductors.

1.6 (a) The magnetic dipole moment $\boldsymbol{\mu}$ of a current loop is defined by

$$\boldsymbol{\mu} = I\mathbf{A},$$

where I is the current, and $\mathbf{A}$ is the area of the loop, the direction of $\mathbf{A}$ being perpendicular to the plane of the loop. A current loop may be represented by a charge e rotating at constant speed in a small circular orbit. Use classical reasoning to show that the magnetic dipole moment of the loop is related to $\mathbf{L}$, the orbital angular momentum of the particle, by

$$\boldsymbol{\mu} = \frac{e}{2m}\mathbf{L},$$

where m is the mass of the particle.

(b) If the magnitude of **L** is $\hbar$ $(= h/2\pi)$, calculate the magnitude of $\boldsymbol{\mu}$ for (i) an electron, and (ii) a proton.

1.7 Calculate the value of the magnetic field required to maintain a stream of protons of energy 1 MeV in a circular orbit of radius 100 mm.

1.8 Neutron diffraction may be used to determine crystal structures.

(a) Estimate a suitable value for the velocity of the neutrons.

(b) Calculate the kinetic energy of the neutron in eV for this velocity.

(c) It is common practice in this type of experiment to select a beam of monoenergetic neutrons from a gas of neutrons at temperature T. Estimate a suitable value for T.

1.9 The most accurate values of the sizes of atomic nuclei come from measurements of electron scattering. Estimate roughly the energies of electrons that provide useful information.

Solutions

1.1 The ionisation energy of hydrogen in the ground state is

$$E_{\text{ion}} = 13.60\ \text{eV} = 2.18 \times 10^{-18}\ \text{J}. \tag{1}$$

The frequency of the radiation that will just ionise the atom is

$$\nu = \frac{E_{\text{ion}}}{h} = 3.29 \times 10^{15}\ \text{Hz}. \tag{2}$$

The wavelength λ and wavenumber $\tilde{\nu}$ of the radiation are

$$\lambda = \frac{c}{\nu} = 9.12 \times 10^{-8}\ \text{m}, \tag{3}$$

$$\tilde{\nu} = \frac{1}{\lambda} = 1.10 \times 10^{7}\ \text{m}^{-1}. \tag{4}$$

1.2 The energy difference between the two levels is

$$\Delta E = \frac{h\nu}{e} = 3.80 \times 10^{-5}\ \text{eV}. \tag{1}$$

1.3 The energy of each photon is

$$E = \frac{hc}{\lambda}, \tag{1}$$

where

$$\lambda = 6.33 \times 10^{-7}\ \text{m}. \tag{2}$$

The power of the laser is

$$P = 1\ \text{mW}. \tag{3}$$

The number of photons emitted per second is

$$n = \frac{P}{E} = \frac{P\lambda}{hc} = 3.19 \times 10^{15}. \tag{4}$$

1.4 (a) The minimum energy E_{min} of the γ-ray photon required for the production of an electron and a positron is equal to the sum of the rest mass energies of the two particles. The mass of the positron is equal to m_e, the mass of the electron. So the required value is

$$E_{\min} = \frac{2m_e c^2}{10^6 e} = 1.02\,\text{MeV}. \quad (1)$$

(b) The frequency ν corresponding to this energy is

$$\nu = \frac{2m_e c^2}{h} = 2.47 \times 10^{20}\ \text{Hz}. \quad (2)$$

1.5 The frequency of oscillation ν of the current is given by

$$h\nu = 2eV. \quad (1)$$

For $V = 1$ volt, the frequency is

$$\nu = 4.84 \times 10^{14}\ \text{Hz}. \quad (2)$$

By measuring the frequency we can deduce the value of the applied voltage from (1). The phenomenon provides a high-precision method of measuring a potential difference – see Solution 8.10, Comment (2) on p. 178.

1.6 (a) Denote the radius of the orbit by a, and the speed of the particle by v. Then the period of revolution is $\tau = 2\pi a/v$. The current due to the rotating charge is

$$I = \frac{e}{\tau} = \frac{ev}{2\pi a}. \quad (1)$$

The magnetic dipole moment is

$$\mu = IA = \frac{ev}{2\pi a}\pi a^2 = \tfrac{1}{2}eva. \quad (2)$$

The orbital angular momentum is

$$L = mav. \quad (3)$$

Therefore

$$\boldsymbol{\mu} = \frac{e}{2m}\mathbf{L}. \quad (4)$$

The vector form follows because, for positive e, the quantities $\boldsymbol{\mu}$ and $\mathbf{L}$ are in the same direction.

(b) For $L = \hbar$, the magnetic dipole moment of a circulating electron is

$$\mu_e = \frac{e\hbar}{2m_e} = 9.28 \times 10^{-24}\ \text{J}\,\text{T}^{-1}, \quad (5)$$

while, for a circulating proton, it is

$$\mu_{\mathrm{p}} = \frac{e\hbar}{2m_{\mathrm{p}}} = 5.05 \times 10^{-27}\,\mathrm{J\,T^{-1}}. \tag{6}$$

Comments

Although the result in (4) has been derived by classical reasoning for the special case of a charge moving in a circular orbit, it is valid for orbital motion in general in quantum mechanics. A particle, such as an electron or a proton, in a stationary state does not move in a definite orbit, nor does it have a definite speed, but it does have a definite orbital angular momentum, the component of which in any direction is of the form $n\hbar$, where n is an integer, positive, negative, or zero. Thus $\hbar$ may be regarded as a natural unit of angular momentum. Since magnetic dipole moment and angular momentum are related by (4), the component of magnetic dipole moment of an electron, due to its orbital motion, has the form $n\mu_{\mathrm{B}}$, where

$$\mu_{\mathrm{B}} = \frac{e\hbar}{2m_{\mathrm{e}}}. \tag{7}$$

Thus μ_{B}, known as the *Bohr magneton*, is the natural unit of magnetic dipole moment for the electron. Similarly the quantity

$$\mu_{\mathrm{N}} = \frac{e\hbar}{2m_{\mathrm{p}}}, \tag{8}$$

known as a *nuclear magneton*, is the natural unit of magnetic dipole moment for the proton.

The simple relation in (4) between magnetic dipole moment and angular momentum does not apply when the effects of the *intrinsic* or *spin* angular momentum of the particle are taken into account. However, it remains true that the magnetic dipole moments of atoms are of the order of Bohr magnetons, while the magnetic dipole moments of the proton, the neutron, and of nuclei in general, are of the order of nuclear magnetons.

1.7 If the velocity of the proton is $\mathbf{v}$, the Lorentz force acting on it, due to the magnetic field $\mathbf{B}$, is $e[\mathbf{v} \times \mathbf{B}]$. The force is perpendicular to the instantaneous direction of motion and to the direction of $\mathbf{B}$. Thus the protons move in a circle, the plane of which is perpendicular to $\mathbf{B}$. Equating the force to the mass times the centripetal acceleration for circular motion, we have

$$Bev = \frac{m_p v^2}{a}, \tag{1}$$

where a is the radius of the circle. Whence

$$B = \frac{m_p v}{ea} = \frac{(2m_p E)^{1/2}}{ea}. \tag{2}$$

Inserting the values of the constants, together with $E = 10^6$ eV $= (10^6 e)$ J, and $a = 0.1$ m, gives

$$B = 1.45 \text{ T}. \tag{3}$$

1.8 (a) To obtain information on the crystal structure, neutrons are diffracted by the crystal in accordance with Bragg's law

$$n\lambda = 2d \sin\theta. \tag{1}$$

(This is the same law that governs the diffraction of X-rays.) In this equation, λ is the wavelength of the neutrons, d is the distance between the planes of diffracting atoms, θ is the glancing angle between the direction of the incident neutrons and the planes of atoms, and n is an integer (usually small). The equation cannot be satisfied unless $\lambda < 2d$. On the other hand, if $\lambda \ll 2d$, θ is inconveniently small. So it is necessary for λ to be of the same order as d, which is of the order of the interatomic spacing in the crystal. Put $\lambda = d = 0.2$ nm (a typical value).

The de Broglie relation between λ and the velocity v of the neutron is

$$\lambda = \frac{h}{m_n v}, \tag{2}$$

where m_n, the mass of the neutron, is 1.675×10^{-27} kg. Thus

$$v = \frac{h}{m_n \lambda} = 2.0 \text{ km s}^{-1}. \tag{3}$$

(b) The kinetic energy of the neutrons is

$$E = \tfrac{1}{2} m_n v^2 = 3.3 \times 10^{-21} \text{ J} = 20 \text{ meV} \tag{4}$$

for the above velocity.

(c) Put $E = k_B T$. Then the above value of E corresponds to $T = 240$ K, which is of the order of room temperature. Such neutrons are readily available in a thermal nuclear reactor; they are termed *thermal neutrons*.

1.9 The electrons scattered by nuclei show diffraction effects characteristic of the radius r of the nucleus, the value of which lies in the range

1–6 fm (1 fm = 10^{-15} m). As in Problem 1.8, measurable effects require that the wavelength λ of the electron should be of the order of r. Thus the momentum p of the electron should satisfy

$$p = \frac{h}{\lambda} \approx \frac{h}{r} = 1.3 \times 10^{-19}\ \mathrm{kg\,m\,s^{-1}}, \qquad (1)$$

for $r = 5$ fm. This value is very much larger than

$$m_e c = 2.7 \times 10^{-22}\ \mathrm{kg\,m\,s^{-1}}, \qquad (2)$$

where m_e is the rest mass of the electron, which shows that the electrons required for the measurements are highly relativistic.

The energy E of the electrons is related to their momentum p by

$$E^2 = m_e^2 c^4 + p^2 c^2. \qquad (3)$$

Since $p \gg m_e c$, we can neglect the first term on the right-hand side of (3). Thus

$$E \approx pc = 4.0 \times 10^{-11}\ \mathrm{J} = 250\ \mathrm{MeV}. \qquad (4)$$

The value obtained for E clearly depends on the value taken for r. If E is in MeV, and λ $(= r)$ is in fm, you may verify that, for the highly relativistic case,

$$E\lambda \approx 10^9 \times \frac{ch}{e} = 1240. \qquad (5)$$

2
Fundamentals

Summary of theory

1 What you need to know

Definitions and properties

Operator, linear operator, functions of operators, commuting and non-commuting operators, eigenfunction, eigenvalue, degeneracy, normalised function, orthogonal functions, Hermitian operator.

2 Postulates of quantum mechanics

(1) The state of a system with n position variables $q_1, q_2, \ldots q_n$ is specified by a state (or wave) function $\psi(q_1, q_2, \ldots q_n)$. All possible information about the system can be derived from this state function. In general, n is three times the number of particles in the system. So for a single particle $n = 3$, and q_1, q_2, q_3 may be the Cartesian coordinates x, y, z, or the spherical polar coordinates r, θ, ϕ, or some other set of coordinates.

(2) To every observable there corresponds a Hermitian operator given by the following rules:

(i) The operator corresponding to the Cartesian position coordinate x is $x \times$ – similarly for the coordinates y and z.

(ii) The operator corresponding to p_x, the x component of linear momentum, is $(\hbar/\mathrm{i})\partial/\partial x$ – similarly for the y and z components.

(iii) To obtain the operator corresponding to any other observable, first write down the classical expression for the observable in terms of x, y, z, p_x, p_y, p_z, and then replace each of these quantities by its corresponding operator according to rules (i) and (ii).

(3) The only possible result which can be obtained when a measurement is made of an observable whose operator is A is an eigenvalue of A.

(4) Let α be an observable whose operator A has a set of eigenfunctions ϕ_j with corresponding eigenvalues a_j. If a large number of

measurements of α are made on a system in the state ψ, then the expectation value of α for the state ψ (i.e. the arithmetic mean of the eigenvalues obtained) is given by

$$\langle A \rangle = \int \psi^* A \psi \,\mathrm{d}\tau, \tag{1}$$

where $\mathrm{d}\tau$ is an element of volume, and the integral is taken over all space.

(5) If the result of a measurement of α is a_r, corresponding to the eigenfunction ϕ_r, then the state function immediately after the measurement is ϕ_r.

This means that in general a measurement changes or disturbs the state of a system. The set of measurements referred to in the 4th postulate are all made on the system in the same state ψ. It is in general necessary to manipulate the system after each measurement to return it to the state ψ before the next measurement is made.

(6) The time variation of the state function of a system is given by

$$\frac{\partial \psi}{\partial t} = \frac{1}{i\hbar} H \psi, \tag{2}$$

where H is the operator formed from the classical Hamiltonian of the system.

Note on Postulate 2 (iii)

If the classical expression for an observable contains a product $\alpha\beta$ whose operators A and B do not commute, then the operator corresponding to $\alpha\beta$ is $\frac{1}{2}(AB + BA)$. Examples of this are rare.

3 Basic deductions from the postulates

(a) Probability of result of measurement

Discrete eigenvalues. Suppose the eigenvalues a_j of A in postulates 4 and 5 are discrete, and that the state function ψ and all the eigenfunctions ϕ_j of A are normalised. To find the probability p_r that the result of a measurement of the observable α is a particular a_r, expand ψ in terms of the ϕ_j, i.e. put

$$\psi = \sum_j c_j \phi_j. \tag{3}$$

Then

$$p_r = |c_r|^2. \tag{4}$$

The coefficient c_r is obtained from

$$c_r = \int \phi_r^* \psi \, \mathrm{d}\tau. \tag{5}$$

If the coefficients c_j are known, a convenient expression for the expectation value is

$$\langle A \rangle = \sum_j p_j a_j = \sum_j |c_j|^2 a_j. \tag{6}$$

Continuous eigenvalues. Let γ be an observable whose operator G has eigenvalues k which form a continuous spectrum, i.e. any real value k is a possible result of a measurement of γ. For simplicity we give the results for the one-dimensional case for a system consisting of a single particle. Denote the eigenfunctions of G by $\phi(k, x)$. The expansion of $\psi(x)$ in terms of $\phi(k, x)$ is not a sum as in (3), but an integration which we write in the form

$$\psi(x) = \int g(k)\phi(k, x) \, \mathrm{d}k. \tag{7}$$

The significance of the function $g(k)$ is similar to that of the coefficients c_j in the previous section. Specifically, $|g(k)|^2 \, \mathrm{d}k$ is equal to the probability that, if the observable γ is measured, the value obtained lies in the range k to $k + \mathrm{d}k$.

Linear momentum. An important example of an observable whose operator has continuous eigenvalues is linear momentum. The eigenfunctions of the operator p_x are

$$\phi(k, x) = c \exp(\mathrm{i}kx), \tag{8}$$

with eigenvalues $\hbar k$; c is a constant. For linear momentum (7) becomes

$$\psi(x) = c \int_{-\infty}^{\infty} g(k) \exp(\mathrm{i}kx) \, \mathrm{d}k. \tag{9}$$

This equation shows that, for linear momentum, $g(k)$ is proportional to the Fourier transform of $\psi(x)$. The function $g(k)$ is obtained from $\psi(x)$ by the relation

$$g(k) \propto \int_{-\infty}^{\infty} \psi(x) \exp(-\mathrm{i}kx) \, \mathrm{d}x, \tag{10}$$

which follows from (9) by the theory of Fourier transforms. An account of this theory will be found in Mathews and Walker (1970), p. 101. The constant of proportionality in (10) is fixed by the requirement that $|g(k)|^2 \, \mathrm{d}k$ be equal to the probability of finding k in the range k to $k + \mathrm{d}k$, and therefore

$$\int_{-\infty}^{\infty} |g(k)|^2 \, \mathrm{d}k = 1. \tag{11}$$

(b) Time variation of state function

Suppose a one-dimensional system has a Hamiltonian H that does not vary with time. Then, if the state function $\psi(0)$ at time $t = 0$ is known, the function at a later time t is given by

$$\psi(t) = \sum_j c_j u_j \exp(-\mathrm{i}E_j t/\hbar), \tag{12}$$

where u_j is an eigenfunction of H with energy E_j, and the coefficients c_j are given by

$$c_j = \int_{-\infty}^{\infty} u_j^* \psi(0)\, \mathrm{d}x. \tag{13}$$

The state function and the eigenfunctions are assumed normalised.

These results are for discrete energy values. If the eigenvalues E_k of the Hamiltonian have a continuous spectrum, with corresponding eigenfunctions u_k, the equations become

$$\psi(x, t) = \int_{-\infty}^{\infty} g(k) u_k \exp(-\mathrm{i}E_k t/\hbar)\, \mathrm{d}k, \tag{14}$$

where

$$g(k) = \int_{-\infty}^{\infty} u_k^* \psi(x, 0)\, \mathrm{d}x. \tag{15}$$

(c) Time variation of expectation value of observable

The time variation of the expectation value of an observable with operator A for a system in the state ψ is

$$\frac{\partial}{\mathrm{d}t}\langle A\rangle = \frac{1}{\mathrm{i}\hbar}\int \psi^*(AH - HA)\psi\, \mathrm{d}\tau = \frac{1}{\mathrm{i}\hbar}\langle AH - HA\rangle. \tag{16}$$

The result assumes that the operator A does not vary with time

4 Gaussian function

The function $f(x) = \exp\{-(x - X)^2/2\sigma^2\}$, specified by the two constants X and σ, is known as a *Gaussian*. It is a function symmetrical about $x = X$, with its maximum value at $x = X$, and which decreases rapidly to zero as $|x - X|$ becomes large compared to σ. It is often convenient to represent the wave function $\psi(x)$ by a Gaussian. The following mathematical results hold for the function. (We put $X = 0$ for simplicity.)

$$\int_{-\infty}^{\infty} \exp(-x^2/2\sigma^2)\,\mathrm{d}x = \sqrt{(2\pi)}\sigma, \tag{17}$$

$$\int_{-\infty}^{\infty} x^2 \exp(-x^2/2\sigma^2)\,\mathrm{d}x = \sqrt{(2\pi)}\sigma^3. \tag{18}$$

The Fourier transform of the function $\exp(-x^2/4\Delta^2)$ is

$$g(k) \propto \int_{-\infty}^{\infty} \exp(-x^2/4\Delta^2)\exp(-\mathrm{i}kx)\,\mathrm{d}x$$

$$\propto \exp(-k^2\Delta^2). \tag{19}$$

The last result, which is derived in Mathews and Walker (1970), p. 106, shows that the Fourier transform of a Gaussian function is itself a Gaussian.

Problems

2.1 ϕ_1 and ϕ_2 are normalised eigenfunctions corresponding to the same eigenvalue. If

$$\int \phi_1^* \phi_2 \, d\tau = d,$$

where d is real, find normalised linear combinations of ϕ_1 and ϕ_2 that are orthogonal to (a) ϕ_1, (b) $\phi_1 + \phi_2$.

2.2 An operator A, corresponding to an observable α, has two normalised eigenfunctions ϕ_1 and ϕ_2, with eigenvalues a_1 and a_2. An operator B, corresponding to an observable β, has normalised eigenfunctions χ_1 and χ_2, with eigenvalues b_1 and b_2. The eigenfunctions are related by

$$\phi_1 = (2\chi_1 + 3\chi_2)/\sqrt{13}, \quad \phi_2 = (3\chi_1 - 2\chi_2)/\sqrt{13}.$$

α is measured and the value a_1 is obtained. If β is then measured and then α again, show that the probability of obtaining a_1 a second time is 97/169.

2.3 A particle moving in one dimension has a state function

$$\psi(x) = \frac{1}{(2\pi\Delta^2)^{1/4}} \exp(-x^2/4\Delta^2),$$

where Δ is a constant. Show the following.

(a) The state function is correctly normalised.

(b) The probability that the particle has linear momentum in the range p to $p + dp$ is $P(p)\,dp$, where

$$P(p) = \left(\frac{2}{\pi}\right)^{1/2} \frac{\Delta}{\hbar} \exp(-2p^2\Delta^2/\hbar^2).$$

(c) The product of the uncertainties in position and momentum has the minimum value allowed by the uncertainty principle.

2.4 Show that, for the wave function

$$\psi(x) = 1/\sqrt{(2a)} \quad |x| < a,$$
$$= 0 \quad |x| > a,$$

the uncertainty in the momentum is infinite.

2.5 For a system of particles of mass m in the state ψ, the formal expression for the particle flux vector (number per unit time through unit area perpendicular to the direction of motion) is

$$\mathbf{F} = \frac{\hbar}{2im}(\psi^* \operatorname{grad} \psi - \psi \operatorname{grad} \psi^*).$$

Show that, for a beam of free particles moving with velocity v in one dimension, the expression gives

$$F = v \times \text{density of particles}.$$

2.6 A free particle travelling in one dimension is represented by the wave

$$\psi = A \exp\{i(kx - \omega t)\}.$$

(a) Calculate the group velocity g of the wave using non-relativistic mechanics, and show that it equals the particle velocity v.

(b) Show that the same result holds for relativistic mechanics.

(c) Show that the relation between the phase velocity u and the group velocity g of the wave is

$$u = g/2 \quad \text{for non-relativistic mechanics,}$$

and

$$u = c^2/g \quad \text{for relativistic mechanics,}$$

where c is the speed of light.

2.7 For a certain system, the operator corresponding to the physical quantity A does not commute with the Hamiltonian. It has eigenvalue a_1 and a_2, corresponding to eigenfunctions

$$\phi_1 = (u_1 + u_2)/\surd 2, \quad \phi_2 = (u_1 - u_2)\surd 2,$$

where u_1 and u_2 are eigenfunctions of the Hamiltonian with eigenvalues E_1 and E_2. If the system is in the state $\psi = \phi_1$ at time $t = 0$, show that the expectation value of A at time t is

$$\langle A \rangle = \frac{a_1 + a_2}{2} + \frac{a_1 - a_2}{2} \cos \frac{(E_1 - E_2)t}{\hbar}.$$

2.8 (a) At time $t = 0$, the state function of a free particle in a one-dimensional system is

$$\psi(x, 0) = c \exp(-x^2/4\Delta_0^2),$$

where c and Δ_0 are constants. Show that Δ_t, the uncertainty in position at time t, is given by

$$\Delta_t^2 = \Delta_0^2 + (\Delta v)^2 t^2,$$

where Δv is the uncertainty in the velocity at $t = 0$.

(b) How does the uncertainty in velocity vary with time?

2.9 The one-dimensional motion of a particle of mass m in a potential $V(x)$ is represented by the state function $\psi(x, t)$.

(a) Prove that the time variations of the expectation values of position and momentum are given by

$$\frac{\mathrm{d}}{\mathrm{d}t}\langle x\rangle = \langle p\rangle/m, \quad \text{and} \quad \frac{\mathrm{d}}{\mathrm{d}t}\langle p\rangle = -\left\langle\frac{\mathrm{d}V}{\mathrm{d}x}\right\rangle.$$

(b) Explain the physical significance of these results.

Solutions

2.1 (a) Let $c_1\phi_1 + c_2\phi_2$ be the linear combination that is orthogonal to ϕ_1. Then, since ϕ_1 is normalised,

$$\int \phi_1^*(c_1\phi_1 + c_2\phi_2)\,\mathrm{d}\tau = c_1 + c_2 d = 0. \tag{1}$$

Therefore $c_1/c_2 = -d$.

The separate values of c_1 and c_2 are obtained from the requirement that $c_1\phi_1 + c_2\phi_2$ be normalised. Take c_1 and c_2 to be real. Then

$$\int (c_1\phi_1 + c_2\phi_2)^*(c_1\phi_1 + c_2\phi_2)\,\mathrm{d}\tau = c_1^2 + c_2^2 + 2dc_1c_2 = 1. \tag{2}$$

Eqs. (1) and (2) give

$$c_1 = \frac{d}{\surd(1-d^2)}, \qquad c_2 = -\frac{1}{\surd(1-d^2)}. \tag{3}$$

(Alternatively the signs of both c_1 and c_2 may be reversed.) So a solution with the required properties is

$$(d\phi_1 - \phi_2)/\surd(1-d^2). \tag{4}$$

Notice that c_1 and c_2 do not have to be real. We may multiply both the expressions in (3) by $\exp(\mathrm{i}\delta)$, where δ is any real number, and the resulting expressions for c_1 and c_2 are a valid solution. This corresponds to the fact that if f is normalised and orthogonal to g, so is $\exp(\mathrm{i}\delta)f$.

(b) Let $c_1\phi_1 + c_2\phi_2$ be normalised, and orthogonal to $\phi_1 + \phi_2$. The orthogonality condition gives

$$\int (\phi_1 + \phi_2)^*(c_1\phi_1 + c_2\phi_2)\,\mathrm{d}\tau = (c_1 + c_2)(1 + d) = 0. \tag{5}$$

Therefore $c_1 = -c_2$. The normalising condition gives (2) as before. From (2) and (5) we find that

$$(\phi_1 - \phi_2)/\surd(2 - 2d) \tag{6}$$

has the required properties.

2.2 After a measurement of α which gives the result a_1, the system is in the state $\psi = \phi_1$. If β is now measured, the probabilities of obtaining the values b_1 and b_2 are given by expanding ϕ_1 in terms of χ_1 and χ_2, the eigenfunctions of the operator B. If

$$\phi_1 = c_1\chi_1 + c_2\chi_2, \tag{1}$$

then the probability of the measurement of β giving b_1 is $|c_1|^2$, and of giving b_2 is $|c_2|^2$. The expansion in (1) is already given in the problem with

$$c_1 = \frac{2}{\sqrt{13}}, \quad c_2 = \frac{3}{\sqrt{13}}. \tag{2}$$

Suppose the result of measuring β is b_1. The system is now in the state χ_1, and the probabilities of obtaining the values a_1 and a_2 when the observable α is measured again are given by expanding χ_1 in terms of ϕ_1 and ϕ_2, the eigenfunctions of the operator A. From (1) and the relation

$$\phi_2 = c_2\chi_1 - c_1\chi_2, \tag{3}$$

we obtain

$$\chi_1 = c_1\phi_1 + c_2\phi_2, \quad \chi_2 = c_2\phi_1 - c_1\phi_2. \tag{4}$$

Thus when the system is in the state χ_1, the probability that a measurement of α gives the result a_1 is c_1^2. So the probability that, from an initial state ϕ_1, successive measurements of β and α yield the values b_1, a_1 is c_1^4. Similarly the probability of the same pair of measurements yielding the values b_2, a_1 is c_2^4.

The total probability of obtaining the value a_1 from the second measurement of α is thus

$$c_1^4 + c_2^4 = \frac{97}{169}. \tag{5}$$

Comment

Note that, if the functions χ_1 and χ_2 are orthogonal, then the functions

$$\phi_1 = c_1\chi_1 + c_2\chi_2 \quad \text{and} \quad \phi_2 = d_1\chi_1 + d_2\chi_2 \tag{6}$$

are orthogonal if

$$c_1 d_1^* + c_2 d_2^* = 0, \tag{7}$$

a condition which is satisfied by the coefficients in (1) and (3). It is part of the postulates of quantum mechanics that the operators corresponding to physical observables are Hermitian, and it is a mathematical result that the eigenfunctions of a Hermitian operator corresponding to different eigenvalues are orthogonal. The present problem has been constructed to satisfy this condition.

2.3 (a) The result follows from (2.17)

$$\int_{-\infty}^{\infty} |\psi|^2 \, dx = \left(\frac{1}{2\pi}\right)^{1/2} \frac{1}{\Delta} \int_{-\infty}^{\infty} \exp(-x^2/2\Delta^2) \, dx = 1. \tag{1}$$

(b) We use the results in Section 3(a), p. 10. The probability that the particle has momentum p to $p + dp$ is

$$P(p) \, dp = |g(k)|^2 \, dk, \tag{2}$$

where $g(k)$ is given by (2.10). Inserting the expression for $\psi(x)$ in (2.10) and using the result (2.19), we obtain

$$\begin{aligned} g(k) &\propto \int_{-\infty}^{\infty} \exp(-x^2/4\Delta^2) \exp(-ikx) \, dx \\ &\propto \exp(-k^2\Delta^2). \end{aligned} \tag{3}$$

Put

$$g(k) = c \exp(-k^2\Delta^2), \tag{4}$$

where c is a constant, the value of which is found from (2.11). We have

$$\begin{aligned} \int_{-\infty}^{\infty} |g(k)|^2 \, dk &= 1 \\ &= c^2 \int_{-\infty}^{\infty} \exp(-2k^2\Delta^2) \, dk \\ &= c^2 \sqrt{(2\pi)} \frac{1}{2\Delta}. \end{aligned} \tag{5}$$

The last step follows from (2.17), with x replaced by k, and $\sigma = 1/(2\Delta)$. From (4) and (5)

$$g(k) = \frac{1}{(2\pi)^{1/4}} (2\Delta)^{1/2} \exp(-k^2\Delta^2). \tag{6}$$

The momentum p is related to the wave number k by $p = \hbar k$. Thus, from (2) and (6),

$$P(p) = |g(k)|^2 \frac{dk}{dp} = \left(\frac{2}{\pi}\right)^{1/2} \frac{\Delta}{\hbar} \exp(-2p^2\Delta^2/\hbar^2). \tag{7}$$

(c) The uncertainties in position and wavenumber are, by definition, the standard deviations of the functions $|\psi(x)|^2$ and $|g(k)|^2$ respectively. We can see from (2.17) and (2.18) that the standard deviation of the function $\exp(-x^2/2\sigma^2)$ is σ. The expressions for $|\psi(x)|^2$ and $|g(k)|^2$ are

$$|\psi(x)|^2 = \frac{1}{(2\pi)^{1/2}} \frac{1}{\Delta} \exp(-x^2/2\Delta^2), \tag{8}$$

$$|g(k)|^2 = \frac{1}{(2\pi)^{1/2}} 2\Delta \exp(-2k^2\Delta^2). \tag{9}$$

$|\psi(x)|^2$ is a Gaussian centred on $x = 0$, with standard deviation Δ. Similarly $|g(k)|^2$ is a Gaussian centred on $k = 0$, with standard deviation $\Delta k = 1/(2\Delta)$. The uncertainty in position Δx is therefore Δ, and the uncertainty in the momentum is

$$\Delta p = \hbar \Delta k = \frac{\hbar}{2\Delta}. \tag{10}$$

Therefore

$$\Delta x \Delta p = \frac{\hbar}{2}. \tag{11}$$

Comment

Eq. (11) is a form of the uncertainty principle for position and momentum. We shall see in Problem 10.1 that the uncertainty principle has the general form

$$\Delta x \Delta p \geqslant \frac{\hbar}{2}. \tag{12}$$

We have proved here that the equality holds for a Gaussian state function, and the solution to Problem 10.1 shows that it holds only for this function.

Notice how the uncertainty principle is built into the theory. Firstly, associating the operator $(\hbar/\mathrm{i})\,\partial/\partial x$ with linear momentum (which may seem somewhat arbitrary) leads directly to the result that a particle with momentum p is in a state represented by the sinusoidal wave $\exp(\mathrm{i}kx)$, where

$$p = \hbar k. \tag{13}$$

The wave number k is equal to $2\pi/\lambda$, where λ is the wavelength of the wave. Thus (13) is equivalent to

$$p = \frac{h}{\lambda}, \tag{14}$$

which is the de Broglie relation between linear momentum and wavelength. The fact that this fundamental relation must be satisfied may be regarded as governing the choice of operator for linear momentum.

Secondly, the postulates lead to the result that the momentum function $g(k)$ is the Fourier transform of the state function $\psi(x)$. Look at (2.9). It says that the function $\psi(x)$ may be represented by the 'sum' of a set of sinusoidal waves $\exp(ikx)$. The 'sum' is of course an integral, that is to say, the set contains an infinite number of waves with continuously varying wavenumber k. The quantity $g(k)$, which is in general complex, gives the amplitude and phase of the wave $\exp(ikx)$. By suitable choice of the function $g(k)$ we can reproduce any function $\psi(x)$ – provided it satisfies certain mild conditions, which all ψs representing actual physical states in fact do.

If we decide we want to know the position of the particle more precisely, we need a more sharply peaked function $\psi(x)$. But the Fourier transform $g(k)$ of such a function contains a greater range of k values, i.e. $g(k)$ becomes broader or less sharply peaked. So, as Δx, the uncertainty in the position, decreases, Δk, and hence Δp, the uncertainty in the momentum, increases. This is clearly seen in the above analysis for the Gaussian function where Δk is proportional to $1/(\Delta x)$.

2.4 We again need to calculate the function $g(k)$. For the wave function of the problem, (2.10) gives

$$\begin{aligned} g(k) &\propto \int_{-\infty}^{\infty} \psi(x) \exp(-ikx)\,dx \\ &\propto \int_{-a}^{a} \exp(-ikx)\,dx \\ &\propto \frac{1}{ik}\{\exp(ika) - \exp(-ika)\} \\ &\propto \frac{1}{k}\sin ka. \end{aligned} \tag{1}$$

The square of the standard deviation Δk of the function $|g(k)|^2$ is proportional to

$$\int_{-\infty}^{\infty} \frac{1}{k^2} \sin^2(ka) k^2\,dk,$$

which is infinite. Therefore, the uncertainty in the momentum $\Delta p = \hbar\Delta k$ is infinite. The uncertainty in position Δx is clearly not zero. (It is readily shown to be $a/\sqrt{3}$.) So the product of the uncertainties in position and momentum is infinite.

Comment

The fact that Δk is infinite for the top-hat wave function comes from the fact that, to reproduce the infinitely sharp edges of this function, Fourier components with very high wavenumbers are needed. The resulting $g(k)$

tends to zero as k tends to $\pm\infty$, but not sufficiently fast to give a finite standard deviation for $|g(k)|^2$. A wave function $\psi(x)$ may be realised physically by passing a stream of particles through a screen with an appropriate transmission function. No physical screen has a transmission function that changes discontinuously in the mathematical sense, which is the required characteristic for the top-hat function, so in practice Δk would not be infinite.

2.5 In one dimension the expression for the flux is

$$F = \frac{\hbar}{2im}\left(\psi^*\frac{\partial\psi}{dx} - \psi\frac{\partial\psi^*}{dx}\right). \tag{1}$$

The wave function of a beam of free particles of density ρ, moving with velocity v is

$$\psi(x) = A\exp(ikx), \tag{2}$$

where

$$\rho = |A|^2, \quad \text{and} \quad k = mv/\hbar. \tag{3}$$

Substituting (2) into (1) gives

$$F = \frac{\hbar k}{m}|A|^2 = v \times \text{density of particles}. \tag{4}$$

This result is to be expected. If the particles have density ρ and velocity v, the number of particles passing through unit area in unit time is equal to the number in a cylinder of unit cross-section and length v, which is equal to $v\rho$.

2.6 (a) The relations between momentum p, wavenumber k, energy E, and angular frequency ω are

$$p = \hbar k, \tag{1}$$

$$E = \frac{p^2}{2m} = \frac{\hbar^2}{2m}k^2 = \hbar\omega. \tag{2}$$

The first equation in (2) is the non-relativistic relation between E and p. Thus

$$\omega = \frac{\hbar}{2m}k^2. \tag{3}$$

The group velocity g is given by

$$g = \frac{d\omega}{dk} = \frac{\hbar k}{m} = \frac{p}{m} = v, \tag{4}$$

where v is the velocity of the particle.

(b) The relativistic relation between E, p, and rest mass m_0 is

$$E^2 = p^2c^2 + (m_0c^2)^2. \tag{5}$$

Therefore

$$\hbar^2\omega^2 = \hbar^2k^2c^2 + (m_0c^2)^2,$$

whence

$$2\hbar^2\omega d\omega = 2\hbar^2 kc^2 dk. \tag{6}$$

As before

$$g = \frac{d\omega}{dk} = \frac{kc^2}{\omega} = \frac{\hbar kc^2}{\hbar\omega} = \frac{pc^2}{mc^2} = v. \tag{7}$$

In this equation we have used the result $E = \hbar\omega = mc^2$, where m is the relativistic mass of the particle.

(c) The phase velocity is given by

$$u = \frac{\omega}{k}. \tag{8}$$

For the non-relativistic case, we have, from (2),

$$u = \frac{\omega}{k} = \frac{\hbar k}{2m} = \frac{g}{2}. \tag{9}$$

For the relativistic case, we have from (7)

$$u = \frac{\omega}{k} = \frac{c^2}{g}. \tag{10}$$

Comments

(1) The group velocity g is the velocity of the wave packet that represents the position of the particle. So it must equal v, the classical velocity of the particle.

(2) Normally, a relativistic result in physics tends to the corresponding non-relativistic result as the velocity involved becomes small compared to the speed of light. This is clearly not the case for the two expressions for the phase velocity u in (9) and (10). The reason is that the expression for the energy in (5) includes the rest-mass term m_0c^2, whereas that in (2)

does not. There is always an arbitrary zero in the definition of the energy of a material particle. The values of the frequency and the phase velocity of a de Broglie wave are similarly arbitrary; they cannot be determined experimentally.

2.7 The state function at time t is obtained by first expressing the state function at $t = 0$ in terms of the eigenfunctions u_j of the Hamiltonian, which in this case is

$$\psi(0) = (u_1 + u_2)/\sqrt{2}, \tag{1}$$

and then multiplying each u_j by the phase factor $\exp(-\mathrm{i}\omega_j t)$, where $\hbar\omega_j = E_j$. Thus

$$\psi(t) = \{u_1 \exp(-\mathrm{i}\omega_1 t) + u_2 \exp(-\mathrm{i}\omega_2 t)\}/\sqrt{2} \tag{2}$$

$$= c_1 \frac{u_1 + u_2}{\sqrt{2}} + c_2 \frac{u_1 - u_2}{\sqrt{2}}. \tag{3}$$

The last expression is an expansion of ψ in terms of the eigenfunctions of A. Then the expectation value of A is

$$\langle A \rangle = |c_1|^2 a_1 + |c_2|^2 a_2. \tag{4}$$

The values of c_1 and c_2 are obtained by equating the coefficients of u_1 and u_2 in (2) and (3), which gives

$$c_1 + c_2 = \exp(-\mathrm{i}\omega_1 t), \quad c_1 - c_2 = \exp(-\mathrm{i}\omega_2 t), \tag{5}$$

whence

$$\begin{aligned} c_1 &= \tfrac{1}{2}\{\exp(-\mathrm{i}\omega_1 t) + \exp(-\mathrm{i}\omega_2 t)\}, \\ c_2 &= \tfrac{1}{2}\{\exp(-\mathrm{i}\omega_1 t) - \exp(-\mathrm{i}\omega_2 t)\}, \\ |c_1|^2 &= \tfrac{1}{2}\{1 + \cos(\omega_1 - \omega_2)t\}, \\ |c_2|^2 &= \tfrac{1}{2}\{1 - \cos(\omega_1 - \omega_2)t\}. \end{aligned} \tag{6}$$

Substituting these expressions, together with $\hbar\omega_j = E_j$, in (4) gives

$$\langle A \rangle = \frac{a_1 + a_2}{2} + \frac{a_1 - a_2}{2} \cos \frac{(E_1 - E_2)t}{\hbar}. \tag{7}$$

2.8 (a) For a free particle of mass m, the eigenfunctions of the Hamiltonian are $u_k(x) = c \exp(\mathrm{i}kx)$, with energy $E_k = \hbar^2 k^2/2m$. (The constant of proportionality c in the expression for $u_k(x)$ is given by the normalisation condition, and is not needed in the present problem.) Since the eigenvalues of the Hamiltonian are continuous, we use (2.14) to obtain $\psi(x, t)$. Thus

$$\psi(x, t) = \int_{-\infty}^{\infty} g(k)\, u_k(x) \exp(-\mathrm{i}E_k t/\hbar)\, \mathrm{d}k \tag{1}$$

$$\propto \int_{-\infty}^{\infty} g(k) \exp(\mathrm{i}kx) \exp(-\mathrm{i}\hbar k^2 t/2m)\, \mathrm{d}k. \tag{2}$$

To obtain $g(k)$ we put $t = 0$ in (2), which gives

$$\psi(x, 0) \propto \exp(-x^2/4\Delta_0^2) \propto \int_{-\infty}^{\infty} g(k) \exp(\mathrm{i}kx)\, \mathrm{d}k, \tag{3}$$

whence, from (2.19),

$$g(k) \propto \exp(-k^2\Delta_0^2). \tag{4}$$

Therefore

$$\psi(x, t) \propto \int_{-\infty}^{\infty} \exp\left\{-k^2\left(\Delta_0^2 + \frac{\mathrm{i}\hbar t}{2m}\right)\right\} \exp(\mathrm{i}kx)\, \mathrm{d}k, \tag{5}$$

i.e. the function $\psi(x, t)$ is (apart from an irrelevant constant of proportionality) the Fourier transform of the function

$$\exp\left\{-k^2\left(\Delta_0^2 + \frac{\mathrm{i}\hbar t}{2m}\right)\right\}.$$

But we know that the Fourier transform of $\exp(-k^2\Delta_0^2)$ is $\exp(-x^2/4\Delta_0^2)$. Therefore the Fourier transform of $\exp(-k^2\Delta_1^2)$, where $\Delta_1^2 = \Delta_0^2 + (\mathrm{i}\hbar t/2m)$, is $\exp(-x^2/4\Delta_1^2)$. Thus

$$\psi(x, t) \propto \exp\left(-\frac{x^2}{4}\frac{1}{a + \mathrm{i}b}\right), \tag{6}$$

where $a = \Delta_0^2$, and $b = \hbar t/2m$.

The uncertainty in position at time t is given by the standard deviation of the function

$$|\psi(x, t)|^2 \propto \exp\left(-\frac{x^2}{4}\frac{1}{a + \mathrm{i}b}\right) \exp\left(-\frac{x^2}{4}\frac{1}{a - \mathrm{i}b}\right)$$

$$= \exp\left(-\frac{x^2}{2}\frac{a}{a^2 + b^2}\right). \tag{7}$$

Since the standard deviation of the Gaussian function $\exp(-x^2/2\sigma^2)$ is σ, the standard deviation Δ_t of $|\psi(x, t)|^2$ is given by

$$\Delta_t^2 = a + \frac{b^2}{a} = \Delta_0^2 + \frac{\hbar^2 t^2}{4m^2\Delta_0^2}. \tag{8}$$

It remains to show that

$$\frac{\hbar}{2m\Delta_0} = \Delta v. \tag{9}$$

At $t = 0$, the probability that the particle has wave number k to $k + \mathrm{d}k$ is $|g(k)|^2 \,\mathrm{d}k$. From (4)

$$|g(k)|^2 \propto \exp(-2k^2\Delta_0^2), \tag{10}$$

which is a Gaussian with standard deviation

$$\Delta_k = \frac{1}{2\Delta_0}. \tag{11}$$

The velocity v is related to the wave number k by $\hbar k = mv$. Therefore the uncertainty in the velocity at $t = 0$ is

$$\Delta v = \frac{\hbar}{m}\Delta_k = \frac{\hbar}{2m\Delta_0}. \tag{12}$$

(b) The uncertainty in the velocity Δv is defined by

$$(\Delta v)^2 = \langle (v - \langle v \rangle)^2 \rangle = \langle v^2 \rangle, \tag{13}$$

since $\langle v \rangle = 0$ for the ψ of the present problem. So the uncertainty in the velocity is related to the expectation value of the kinetic energy T by

$$\langle T \rangle = \tfrac{1}{2}m\langle v^2 \rangle = \tfrac{1}{2}m(\Delta v)^2. \tag{14}$$

For a free particle, $\langle T \rangle$ is constant in time, and therefore so is Δv. (For a different wave function $\psi(x, 0)$, $\langle v \rangle$ may not be zero, but it would still remain constant for a free particle. So the conclusion that Δv is constant would still hold.)

The same result may be derived formally. The probability that the particle has wave number k to $k + \mathrm{d}k$ at time t is $|g(k, t)|^2 \,\mathrm{d}k$, where $g(k, t)$ is given by

$$\psi(x, t) = \int_{-\infty}^{\infty} g(k, t) u_k(x) \,\mathrm{d}k. \tag{15}$$

Comparing (1) and (15) we see that

$$g(k, t) = g(k) \exp(-\mathrm{i}E_k t/\hbar). \tag{16}$$

Therefore

$$|g(k, t)|^2 = |g(k)|^2. \tag{17}$$

So the uncertainty in k, and hence in the velocity, remains constant in time.

Comment

The result $\Delta_t^2 = \Delta_0^2 + (\Delta v)^2 t^2$ has a simple interpretation. At time $t = 0$, the uncertainty in the position is Δ_0, and the uncertainty in the velocity is Δv. At a later time t, there are two contributions to the uncertainty in position, namely, the original Δ_0, and $(\Delta v)t$ due to the uncertainty in the velocity. These contributions are uncorrelated, and the result of the problem shows that they are combined to give the resultant uncertainty in position at time t in exactly the same way as we combine two uncorrelated errors in a classical measurement.

2.9 (a) We use the results that the expectation value of an observable with operator A for a system in the state ψ is given by

$$\langle A \rangle = \int \psi^* A \psi \, \mathrm{d}x, \tag{1}$$

and that, provided A does not vary with time, the time variation of $\langle A \rangle$ is given by

$$\frac{\mathrm{d}}{\mathrm{d}t}\langle A \rangle = \frac{1}{\mathrm{i}\hbar}\langle AH - HA \rangle, \tag{2}$$

where H is the Hamiltonian operator. Therefore, for the operator corresponding to position, which is simply (x times), we have

$$\frac{\mathrm{d}}{\mathrm{d}t}\langle x \rangle = \frac{1}{\mathrm{i}\hbar}\langle xH - Hx \rangle. \tag{3}$$

The Hamiltonian operator has the form

$$H = \frac{p^2}{2m} + V(x). \tag{4}$$

The operator x commutes with $V(x)$. Therefore

$$xH - Hx = \frac{1}{2m}(xp^2 - p^2 x). \tag{5}$$

The right-hand side is evaluated starting with the commutation relation

$$px - xp = \frac{\hbar}{\mathrm{i}}, \tag{6}$$

and performing some operator algebra, which is the same as ordinary algebra, except that the order of multiplication for operators must be preserved. Multiplying (6), first by p on the left and then by p on the right, and then adding the results gives

$$p^2x - pxp = \frac{\hbar}{\mathrm{i}}p,$$

$$pxp - xp^2 = \frac{\hbar}{\mathrm{i}}p,$$

whence

$$p^2x - xp^2 = \frac{2\hbar}{\mathrm{i}}p. \tag{7}$$

From (3), (5), and (7)

$$\frac{\mathrm{d}}{\mathrm{d}t}\langle x\rangle = \langle p\rangle/m. \tag{8}$$

To prove the second relation we put $A = p$ in (2). The operator corresponding to p is $(\hbar/\mathrm{i})\,\mathrm{d}/\mathrm{d}x$. Since p commutes with any power of itself, we have

$$\frac{\mathrm{d}}{\mathrm{d}t}\langle p\rangle = \frac{1}{\mathrm{i}\hbar}\langle pH - Hp\rangle = \frac{1}{\mathrm{i}\hbar}\langle pV - Vp\rangle \tag{9}$$

$$= -\int \psi^*\left(\frac{\mathrm{d}}{\mathrm{d}x}V - V\frac{\mathrm{d}}{\mathrm{d}x}\right)\psi\,\mathrm{d}\tau$$

$$= -\int \psi^*\frac{\mathrm{d}V}{\mathrm{d}x}\psi\,\mathrm{d}\tau \tag{10}$$

$$= -\left\langle\frac{\mathrm{d}V}{\mathrm{d}x}\right\rangle. \tag{11}$$

At line (10) we have used the result

$$\left(\frac{\mathrm{d}}{\mathrm{d}x}V - V\frac{\mathrm{d}}{\mathrm{d}x}\right)\psi = \frac{\mathrm{d}}{\mathrm{d}x}(V\psi) - V\frac{\mathrm{d}\psi}{\mathrm{d}x}$$

$$= \frac{\mathrm{d}V}{\mathrm{d}x}\psi + V\frac{\mathrm{d}\psi}{\mathrm{d}x} - V\frac{\mathrm{d}\psi}{\mathrm{d}x} = \frac{\mathrm{d}V}{\mathrm{d}x}\psi. \tag{12}$$

(b) The significance of (8) is that the velocity of a particle may be calculated in two equivalent ways. Suppose we have a particle represented by a well-localised wave packet. The wave packet at successive times t_1 and t_2 will appear as shown in Fig. 2.1. The expectation values $\langle x\rangle_1$ and $\langle x\rangle_2$ for the two wave functions correspond classically to the positions of the particle at the two times. So the velocity is given by

$$\frac{\langle x\rangle_2 - \langle x\rangle_1}{t_2 - t_1},$$

which, in the limit as $t_2 - t_1$ tends to zero, is $\mathrm{d}\langle x\rangle/\mathrm{d}t$. Alternatively,

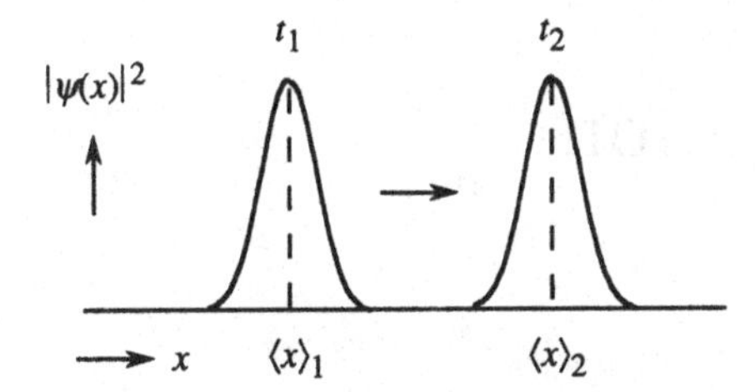

Fig. 2.1. *The velocity of a particle may be obtained either from the motion of its wave packet, as shown in the figure, or from the expectation value of its momentum at a single time.*

from the wave function ψ at a single time we can calculate the expectation value of the momentum from

$$\langle p\rangle = \int \psi^* p \psi \mathrm{d}\tau. \tag{13}$$

Dividing $\langle p\rangle$ by m gives the expectation value of the velocity, which must of course equal $\mathrm{d}\langle x\rangle/\mathrm{d}t$, as (8) shows.

Since $-\mathrm{d}V/\mathrm{d}x$ is the operator corresponding to the force, we have in (11) the result that the time derivative of the expectation value of the momentum is equal to the expectation value of the force. If the particle is well-localised and represented by a narrow wave function, the expectation values of the operators correspond to the classical limits of the observables. In other words, (11) corresponds to Newton's second law, and is an example of the *correspondence principle*. This says that quantum mechanics must produce the same result as classical mechanics for a system in which the particles can be represented by well-localised state functions. The result in (11) is known as *Ehrenfest's theorem*.

3
Schrödinger equation

Summary of theory

1 Time-independent Schrödinger equation

For a one-dimensional system consisting of a particle of mass m moving in a potential $V(x)$, the Hamiltonian is

$$H = \frac{1}{2m}p_x^2 + V(x), \tag{1}$$

where p_x is the momentum of the particle. The first term is the kinetic energy and the second term is the potential energy. The Hamiltonian operator is formed from (1) by replacing p_x by $(\hbar/\mathrm{i})\,\mathrm{d}/\mathrm{d}x$, and is thus

$$H = -\frac{\hbar^2}{2m}\frac{\mathrm{d}^2}{\mathrm{d}x^2} + V. \tag{2}$$

If $u(x)$ is an eigenfunction of the Hamiltonian with eigenvalue E, which in this case is the energy of the system, then

$$Hu = Eu. \tag{3}$$

Inserting the expression for the operator in (2) gives

$$\left(-\frac{\hbar^2}{2m}\frac{\mathrm{d}^2}{\mathrm{d}x^2} + V\right)u = Eu, \tag{4}$$

i.e.

$$\frac{\mathrm{d}^2u}{\mathrm{d}x^2} + \frac{2m}{\hbar^2}(E - V)u = 0. \tag{5}$$

This is the time-independent Schrödinger equation in one dimension.

In three dimensions

$$H = \frac{1}{2m}(p_x^2 + p_y^2 + p_z^2) + V(x, y, z), \tag{6}$$

and the Schrödinger equation is

$$\nabla^2 u + \frac{2m}{\hbar^2}(E - V)u = 0, \tag{7}$$

where u is a function of x, y, z, and ∇^2 is the operator

$$\nabla^2 = \frac{\partial^2}{\partial x^2} + \frac{\partial^2}{\partial y^2} + \frac{\partial^2}{\partial z^2}. \tag{8}$$

2 Continuity conditions for the eigenfunctions

The eigenfunction u is continuous everywhere. The derivatives of u, i.e. $\partial u/\partial x$, $\partial u/\partial y$, $\partial u/\partial z$ are continuous everywhere, except where the potential function has an infinite discontinuity. This cannot happen in an actual physical situation, but it is sometimes convenient to assume it in a theoretical problem.

3 Parity

Consider a one-dimensional function $f(x)$. If $f(x) = f(-x)$, the function is said to have *even* parity, and if $f(x) = -f(-x)$, it is said to have *odd* parity. A function that satisfies neither of these conditions is said to have *mixed* parity; it can always be expressed as a sum of two functions, one with even, and one with odd parity.

If the potential $V(x)$ has even parity, then, if the energy E is non-degenerate, the solution $u(x)$ of the Schrödinger equation has either even or odd parity. If E is degenerate, a solution may have mixed parity, in which case the even and odd parts are separately solutions with the same energy E. So we can always find solutions that have a definite parity.

The same definitions apply in three dimensions when the point $\mathbf{r}$ goes to $-\mathbf{r}$, i.e. if $f(x, y, z) = f(-x, -y, -z)$, the function has even parity, and so on. If the three-dimensional potential $V(\mathbf{r})$ has even parity, the above statements hold for the parity of the solutions of the three-dimensional Schrödinger equation.

Problems

3.1 A particle of mass m moves in a one-dimensional potential which is zero in the region $|x| < a$, and infinite outside this region.

(a) Derive expressions for the normalised solutions of the Schrödinger equation and the corresponding energies.

(b) Sketch the form of the wave functions for the four lowest energies.

(c) What are the parities of the wave functions?

[Notation: $k^2 = 2mE/\hbar^2$.]

3.2 A particle of mass m moves freely in a rectangular box with impenetrable walls.

(a) If the dimensions of the box are $2a_x$, $2a_y$, $2a_z$, derive expressions for the solutions of the Schrödinger equation and the corresponding energies.

(b) What are the parities of the wave functions?

(c) If $a_x = a_y = a_z = a$, what are the degeneracies of the two lowest values of the energy?

3.3 A particle of mass m moves in a one-dimensional potential given by

$$V(x) = -W \quad |x| < a,$$
$$= 0 \quad |x| > a,$$

where W is a positive constant – Fig. 3.1.

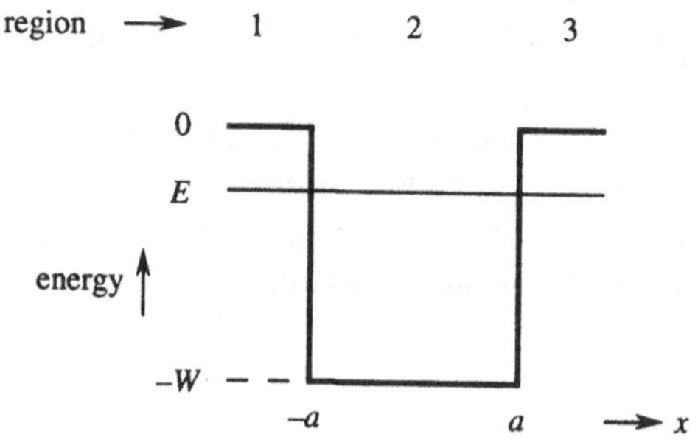

Fig. 3.1. *Potential function and energy for Problem 3.3.*

(a) Show that for a bound state the energy E satisfies the relations

$$\tan ja = \gamma/j \quad \text{for a state of even parity,}$$

$$\cot ja = -\gamma/j \quad \text{for a state of odd parity,}$$

where

$$j^2 = 2m(E + W)/\hbar^2, \quad \text{and} \quad \gamma^2 = -2mE/\hbar^2.$$

(b) Devise a graphical method for determining the values of E, and show that, whatever the values of W and a, there is at least one bound state.

(c) If the particle is an electron and $W = 10\,\text{eV}$, $a = 4 \times 10^{-10}$ m, how many bound states are there? Sketch the form of the wave function for the two lowest bound states.

3.4 A stream of particles of mass m and energy E encounter a potential step of height $W(< E)$ – Fig. 3.2.

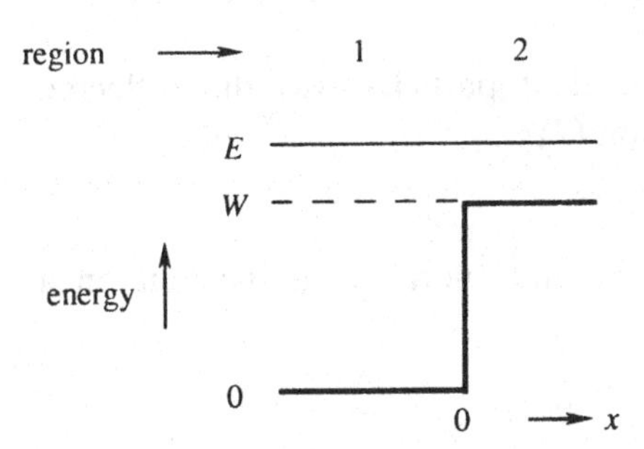

Fig. 3.2. *Potential function and energy for Problem 3.4.*

(a) Show that the fraction reflected is

$$\mathcal{R} = \left(\frac{1 - \mu}{1 + \mu}\right)^2,$$

where

$$\mu = \frac{j}{k}, \quad k^2 = \frac{2m}{\hbar^2}E, \quad j^2 = \frac{2m}{\hbar^2}(E - W).$$

(b) Show that the sum of the fluxes of the reflected and transmitted particles is equal to the flux of the incident particles.

3.5 Consider the last problem for the case $E < W$ – Fig. 3.3. Put

$$k^2 = \frac{2m}{\hbar^2}E, \quad \gamma^2 = \frac{2m}{\hbar^2}(W - E).$$

If the incident particles are represented by $\exp(ikx)$, show

(a) that the reflected particles are represented by $\exp\{-i(kx + 2\theta)\}$, where $\tan\theta = \gamma/k$,

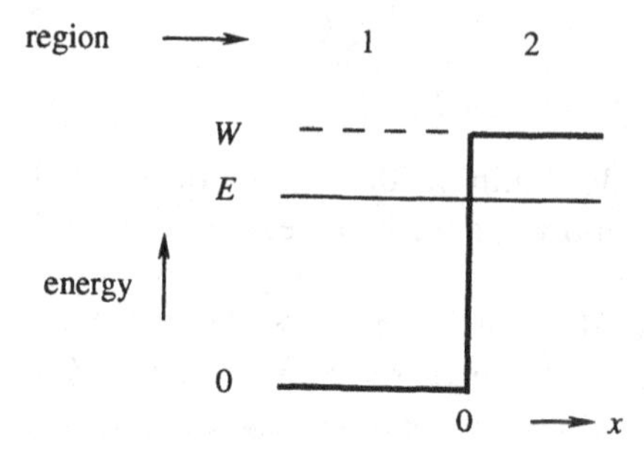

Fig. 3.3. *Potential function and energy for Problem 3.5.*

(b) that the amplitude of the wave function is $2\cos(kx+\theta)$ in region (1), and $2\cos\theta\exp(-\gamma x)$ in region (2).

(c) What is the flux of (i) the incident particles, (ii) the reflected particles, and (iii) the particles in region (2)?
[Notation: $\rho = \gamma/k$.]

3.6 A stream of particles of mass m and energy E is incident on a potential barrier given by

$$V(x) = 0 \qquad x < 0 \quad \text{and} \quad x > a,$$
$$= W \qquad 0 < x < a,$$

where $W > E$ – Fig. 3.4. Show that the fraction transmitted to region (3) is

$$\mathcal{T} = \left\{1 + \frac{W^2}{4E(W-E)}\sinh^2\gamma a\right\}^{-1},$$

where

$$\gamma^2 = 2m(W-E)/\hbar^2.$$

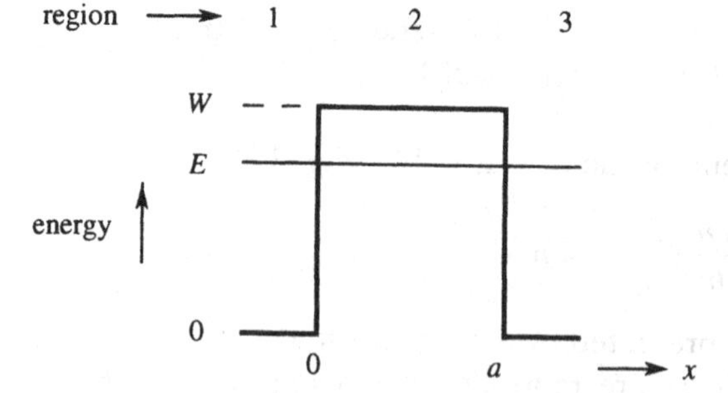

Fig. 3.4. *Potential function and energy for Problem 3.6.*

[Notation: $u_1 = \exp(ikx) + R\exp(-ikx)$, $u_2 = A\exp(\gamma x) + B\exp(-\gamma x)$,

$u_3 = T\exp(ikx)$, $k^2 = 2mE/\hbar^2$, $\rho = \gamma/k$.]

3.7 If the potential in the last problem is a δ function, i.e. $W \to \infty$ and $a \to 0$ in such a way that Wa is equal to a constant b, show that

$$\mathcal{T} = \left(1 + \frac{mb^2}{2\hbar^2 E}\right)^{-1}.$$

Solutions

3.1 (a) For $-a < x < a$, the Schrödinger equation is

$$\frac{d^2u}{dx^2} + k^2u = 0. \tag{1}$$

Solutions of this equation are $\cos kx$ and $\sin kx$. (See Comment 1 below.) Now $u(x)$ is continuous, and since it must be zero for $|x| > a$, we have

$$u(x) = 0 \quad \text{for} \quad x = \pm a.$$

This condition is satisfied if

$$k = n\frac{\pi}{2a}, \tag{2}$$

where n is an odd integer for the cosine solutions, and an even integer (not zero) for the sine solutions. The function with $n = 0$ is ruled out, because it corresponds to $u(x) = 0$ everywhere. This would mean that there is zero probability of finding the particle anywhere, which is not a permissible state.

Let

$$u(x) = A\cos\left(\frac{n\pi x}{2a}\right). \tag{3}$$

The constant A is fixed by the normalisation condition $\int_{-\infty}^{\infty} |u(x)|^2\,dx = 1$. The left-hand side is equal to

$$A^2\int_{-a}^{a} \cos^2\left(\frac{n\pi x}{2a}\right) dx = A^2\frac{2a}{n\pi}\int_{-n\pi/2}^{n\pi/2} \cos^2\theta\,d\theta = A^2a = 1. \tag{4}$$

Similar algebra shows that the normalising constant has the same value for the sine solution.

The required solutions are therefore

$$\begin{aligned} |x| < a \quad & u(x) = \frac{1}{\sqrt{a}}\cos\left(\frac{n\pi x}{2a}\right) \quad n \text{ odd}, \\ & u(x) = \frac{1}{\sqrt{a}}\sin\left(\frac{n\pi x}{2a}\right) \quad n \text{ even}, \\ |x| > a \quad & u(x) = 0. \end{aligned} \tag{5}$$

The energy is given by

$$E = \frac{\hbar^2k^2}{2m} = n^2\frac{\pi^2\hbar^2}{8ma^2}. \tag{6}$$

(b) The wave functions for $n = 1, 2, 3, 4$ are shown in Fig. 3.5. We see that the function has n half-wavelengths in the length $2a$, i.e.

$$n\frac{\lambda}{2} = 2a, \tag{7}$$

where λ is the wavelength. Therefore

$$k = \frac{2\pi}{\lambda} = n\frac{\pi}{2a}, \tag{8}$$

which is a quick way of obtaining the result in (2).

(c) The cosine solutions have even parity, and the sine solutions have odd parity.

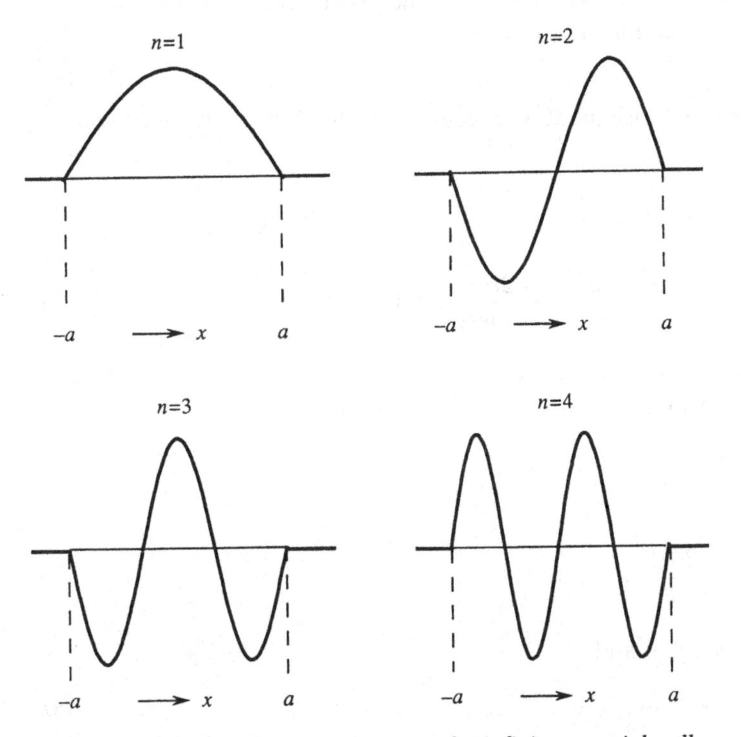

Fig. 3.5. *First four energy eigenfunctions for infinite potential well.*

Comments

(1) Since the potential $V(x)$ has even parity, and the energy values are non-degenerate, the wave function $u(x)$ must have a definite parity, either even or odd. Therefore the solutions must be either cosine or sine functions, and not a linear combination of the two. If we had not made

use of this result, we would have started with a solution of the form

$$u(x) = A \cos kx + B \sin kx, \tag{9}$$

where A and B are constants. Application of the boundary conditions $u(a) = 0$, and $u(-a) = 0$ then gives two equations, which can only both be satisfied if either $B = 0$, and $\cos ka = 0$, or $A = 0$, and $\sin ka = 0$, which are the same results as before.

(2) We could have chosen the origin of x so that the zero-potential region extended from $x = 0$ to $x = 2a$. In that case $V(x)$ would not have even parity, and the wave functions would not have a definite parity. Of course, nothing would be changed physically in the problem. The wave functions would still look the same as in Fig. 3.5, and the energy values would be the same. This shows that the parity property of the wave function depends on the choice of origin.

3.2 (a) Take the origin at the centre of the box. The Schrödinger equation for

$$|x| < a_x, \quad |y| < a_y, \quad |z| < a_z,$$

is

$$\frac{\partial^2 u}{\partial x^2} + \frac{\partial^2 u}{\partial y^2} + \frac{\partial^2 u}{\partial z^2} + \frac{2mE}{\hbar^2} u = 0. \tag{1}$$

The boundary conditions are $u = 0$ at the walls of the box. It is readily verified that the solutions of (1) may be expressed as

$$u = u_x(x) u_y(y) u_z(z), \tag{2}$$

where u_x satisfies

$$\frac{\mathrm{d}^2 u_x}{\mathrm{d}x^2} + \frac{2mE_x}{\hbar^2} u_x = 0, \tag{3}$$

(similarly for u_y, u_z), and

$$E = E_x + E_y + E_z. \tag{4}$$

From the results of Problem 3.1

$$u_x = \frac{1}{\sqrt{a_x}} \cos\left(\frac{n_x \pi x}{2a_x}\right), \quad \text{or} \quad u_x = \frac{1}{\sqrt{a_x}} \sin\left(\frac{n_x \pi x}{2a_x}\right), \tag{5}$$

and n_x is an odd integer for the cosine function and an even integer (not zero) for the sine function; u_y and u_z have similar forms.

The energy is given by

$$E = \frac{\pi^2\hbar^2}{8m}\left(\frac{n_x^2}{a_x^2} + \frac{n_y^2}{a_y^2} + \frac{n_z^2}{a_z^2}\right), \tag{6}$$

where n_x, n_y, n_z is a trio of integers (none zero).

(b) The parity of the function u_x is even for a cosine and odd for a sine function. The parity of u is obtained by multiplying the parities of u_x, u_y, and u_z treating an even parity as $+1$ and an odd parity as -1. Thus the parity of u is even if the three functions u_x, u_y, u_z have even parity, or if one is even and the other two are odd. Therefore the parity of u is

$$\text{even} \quad \text{if} \quad n_x + n_y + n_z \quad \text{is odd,}$$

and

$$\text{odd} \quad \text{if} \quad n_x + n_y + n_z \quad \text{is even.}$$

(c) A state, i.e. an eigenfunction, is specified in this problem by giving the trio of integers n_x, n_y, n_z. The energy is

$$E = \frac{\pi^2\hbar^2}{8ma^2}(n_x^2 + n_y^2 + n_z^2). \tag{7}$$

The state of lowest energy is (1, 1, 1), with energy

$$E = 3\frac{\pi^2\hbar^2}{8ma^2}. \tag{8}$$

There is only one state with this energy, so it is non-degenerate.

There are three states corresponding to the next energy. They are (2, 1, 1), (1, 2, 1), (1, 1, 2). The energy is therefore

$$E = 6\frac{\pi^2\hbar^2}{8ma^2}, \tag{9}$$

and it is three-fold degenerate.

3.3 (a) The Schrödinger equation is

$$\frac{d^2u}{dx^2} + \frac{2m}{\hbar^2}(E - V)u = 0. \tag{1}$$

For a bound state we require solutions with negative E. Eq. (1) becomes (see Fig. 3.1)

$$u'' + j^2u = 0, \quad j^2 = \frac{2m}{\hbar^2}(W + E) \quad \text{region (2),} \tag{2}$$

$$u'' - \gamma^2u = 0, \quad \gamma^2 = -\frac{2m}{\hbar^2}E \quad \text{regions (1) and (3).} \tag{3}$$

Since E is negative, γ is real, and we take it to be positive. The potential has even parity. The solutions therefore have either even or odd parity (or, if of mixed parity, the even and odd parts are separately solutions). We may therefore consider the even and odd parity solutions separately, and it simplifies the algebra to do so.

The form of a solution with even parity is

$$u = A\cos jx, \qquad \text{region (2)}, \tag{4}$$

$$u = B\exp(-\gamma x), \quad \text{region (3)}. \tag{5}$$

The function $\exp(\gamma x)$ is also a solution in region (3), but we reject it immediately because it tends to infinity as x tends to infinity. The boundary conditions are that both u and u' are continuous at $x = a$. Therefore

$$A\cos ja = B\exp(-\gamma a), \tag{6}$$

$$-jA\sin ja = -\gamma B\exp(-\gamma a). \tag{7}$$

Dividing one equation by the other gives

$$\tan ja = \frac{\gamma}{j}. \tag{8}$$

The solution in region (1) is $u = B\exp(\gamma x)$. Applying the boundary conditions at $x = -a$ gives the same pair of equations as before.

The solutions of odd parity have the form

$$u = A\sin jx \qquad \text{region (2)}, \tag{9}$$

$$u = B\exp(-\gamma x) \quad \text{region (3)}. \tag{10}$$

Applying the same boundary conditions as before and dividing the two resulting equations gives

$$\cot ja = -\frac{\gamma}{j}. \tag{11}$$

(b) The quantity that varies in the solution to the problem is the energy E of the particle. We are looking for solutions with negative energy. A value of E less than $-W$ is not physically possible, because the kinetic energy would be everywhere negative. Therefore E may vary from $-W$ to zero. As E varies, both j and γ vary. When (8) is satisfied, we have a solution of positive parity, and when (11) is satisfied, one of odd parity.

Since these equations are transcendental, exact solutions can only be found numerically. To obtain the solutions graphically, we plot the functions

$$\tan ja, \quad -\cot ja, \quad \frac{\gamma}{j}$$

against ja – Fig. 3.6. The general form of γ/j is obtained from the relation

$$\frac{\gamma}{j} = \left(\frac{-E}{W+E}\right)^{1/2}. \tag{12}$$

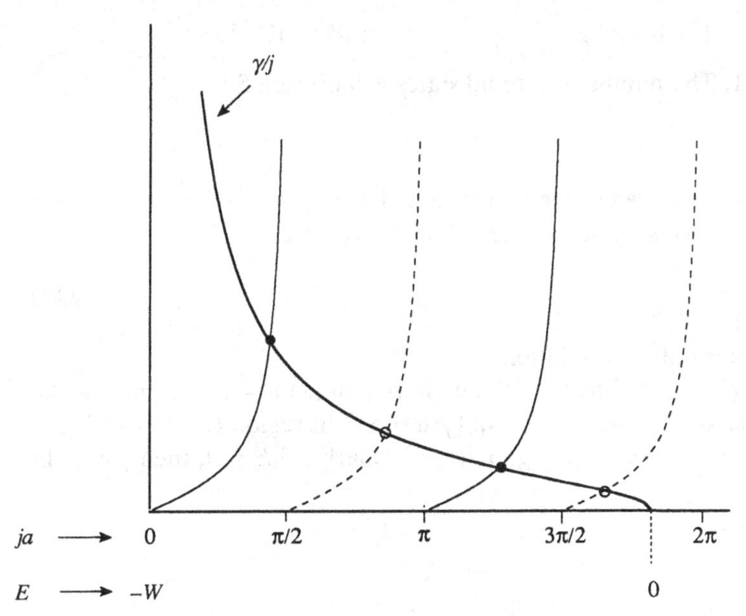

Fig. 3.6. *Graphical solutions of eqs. (8) and (11) in Problem 3.3.*
—— tan *ja* ● *even-parity solution*
---- −cot *ja* ○ *odd-parity solution*

When $E = -W$, γ/j is infinite. As E increases, γ/j decreases monotonically until it reaches the value zero for $E = 0$. Whenever the γ/j curve crosses a tan or −cot curve in Fig. 3.6, we have a solution to the Schrödinger equation which satisfies all the necessary conditions.

Whatever the values of W and a, the γ/j curve must cross the first tan curve. So there must be at least one solution.

(c) The γ/j curve can cross a particular tan or −cot curve only once. The total number of crossing points depends on the value of ja for $E = 0$. When $E = 0$,

$$j = (2mW)^{1/2}/\hbar. \tag{13}$$

Put

$$(2mW)^{1/2}\frac{a}{\hbar} = p\frac{\pi}{2}. \tag{14}$$

Then the number of solutions is the next integer greater than p. The values

$$W = 10\,\text{eV} = 1.6 \times 10^{-18}\,\text{J}, \quad a = 4 \times 10^{-10}\,\text{m},$$

$$m = 9 \times 10^{-31}\,\text{kg}, \quad \hbar = 1.05 \times 10^{-34}\,\text{J s},$$

give $p = 4.1$. The number of bound states is therefore 5.

Comment

It is interesting to reach the conclusion that there must be at least one bound state by looking at the form of u. Assume that

$$(2mW)^{1/2}\frac{a}{\hbar} < \frac{\pi}{2}, \tag{15}$$

so that there is only one solution.

Suppose $E = -W$. Then $j = 0$, and in region (2) $u = A$. We may make u continuous by putting $u = A\exp\{\gamma(a - x)\}$ in region (3), but u' is not continuous at $x = \pm a$ – see Fig. 3.7(a). Similarly, if $E = 0$, then $\gamma = 0$. In

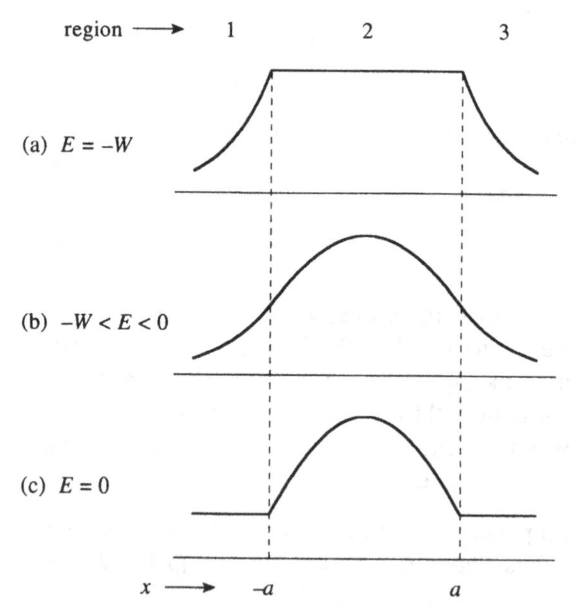

Fig. 3.7. *The form of the wave function $u(x)$ for different values of the energy E in Problem 3.3. Only the one in (b) with continuous slope throughout the range of x is acceptable, and corresponds to the solution for the lowest energy level.*

that case, u is a constant in regions (1) and (3), and if we make u continuous the result is as shown in Fig. 3.7(c). Again the slope at $x = \pm a$ is discontinuous.

In the first case, as x increases, the slope of u decreases at $x = a$, and in the second case it increases. Suppose we let E rise from its value $-W$. As it increases, j starts to increase, and the slope of $A \cos jx$ starts to decrease at $x = a$. At the same time γ starts to decrease, and the slope of $\exp(-\gamma x)$ starts to become less negative. Somewhere between $E = -W$ and $E = 0$ the two slopes will become equal, and u will appear as shown in Fig. 3.7(b).

The forms of the solutions for the next two energy levels are shown in Fig. 3.8.

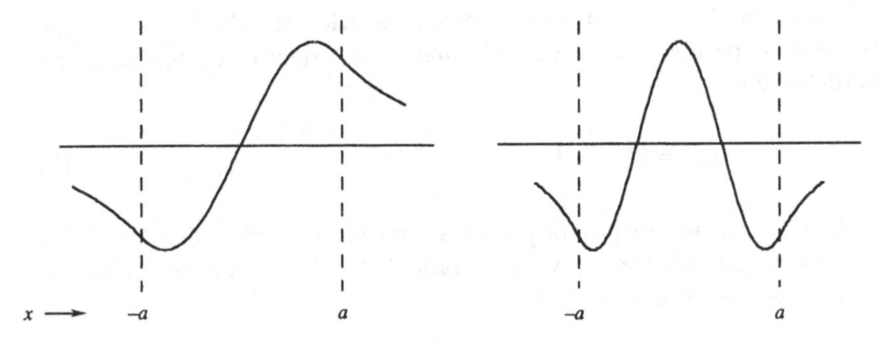

Fig. 3.8. *The form of the wave function $u(x)$ for the second and third lowest energy levels in Problem 3.3.*

3.4 (a) The Schrödinger equation is

$$u'' + k^2 u = 0 \quad \text{region (1)}, \quad u'' + j^2 u = 0 \quad \text{region (2)}. \tag{1}$$

The incident particles in region (1) may be represented by the function $\exp(\mathrm{i}kx)$. (For simplicity we put the multiplying constant equal to unity. Its value does not affect the reasoning in the solution.) The required solution has the form

$$\exp(\mathrm{i}kx) + R\exp(-\mathrm{i}kx) \quad \text{region (1)}, \tag{2}$$

$$T\exp(\mathrm{i}jx) \quad \text{region (2)}, \tag{3}$$

where R and T are constants. The time-dependent factor multiplying all the terms is $\exp(-\mathrm{i}\omega t)$, where $\omega = E/\hbar$. The second term in (2) is thus a wave travelling to the left, representing the particles reflected at the potential step, and the term in (3) is a wave travelling to the right, representing the transmitted particles. The densities of the reflected and transmitted beams are $|R|^2$ and $|T|^2$.

The coefficients R and T are obtained from the continuity conditions. Since u and u' are continuous at $x = 0$, we have

$$1 + R = T, \tag{4}$$

$$k(1 - R) = jT, \tag{5}$$

whence

$$R = \frac{1-\mu}{1+\mu}, \quad T = \frac{2}{1+\mu}, \quad \mu = \frac{j}{k}. \tag{6}$$

The density of the incident particles is unity. Therefore, since the velocities of the incident and reflected particles are the same, $\mathcal{R}$, the fraction of particles reflected, is equal to the density of the reflected particles. So

$$\mathcal{R} = |R|^2 = \left(\frac{1-\mu}{1+\mu}\right)^2. \tag{7}$$

We see that the fraction of particles reflected and transmitted depends only on μ, and therefore only on the ratio E/W. The reflection coefficient $\mathcal{R}$, is plotted against E/W in Fig. 3.9.

(b) From (2.5.4) the flux of particles for a wave $A\exp(ikx)$ is equal to

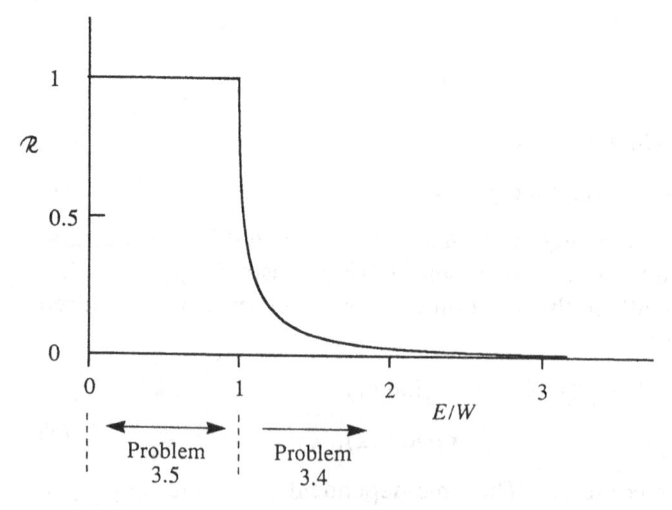

Fig. 3.9. *$\mathcal{R}$, the fraction of particles reflected at a potential step, as a function of E/W in Problems 3.4 and 3.5.*

$(\hbar k/m)|A|^2$, where A is the amplitude of the wave. Without the constant factor $\hbar/m$, these quantities are

$$k, \quad kR^2, \quad jT^2,$$

for the incident, reflected and transmitted waves. From (6)

$$\begin{aligned} kR^2 + jT^2 &= k\left\{\left(\frac{1-\mu}{1+\mu}\right)^2 + \mu\left(\frac{2}{1+\mu}\right)^2\right\} \\ &= k\left\{\frac{1-2\mu+\mu^2+4\mu}{(1+\mu)^2}\right\} \\ &= k. \end{aligned} \tag{8}$$

3.5 (a) The wave function is

$$u_1 = \exp(\mathrm{i}kx) + R\exp(-\mathrm{i}kx) \quad \text{region (1)}, \tag{1}$$

$$u_2 = T\exp(-\gamma x) \quad \text{region (2)}. \tag{2}$$

The solution of the form $\exp(\gamma x)$ is excluded in region (2), because it tends to infinity as x tends to infinity.

u and u' are continuous at $x = 0$. Therefore

$$1 + R = T, \tag{3}$$

$$\mathrm{i}k(1 - R) = -\gamma T, \tag{4}$$

whence

$$R = \frac{1-\mathrm{i}\rho}{1+\mathrm{i}\rho}, \quad T = \frac{2}{1+\mathrm{i}\rho}, \quad \rho = \frac{\gamma}{k}. \tag{5}$$

Put

$$\rho = \tan\theta. \tag{6}$$

Then

$$R = \frac{1-\mathrm{i}\tan\theta}{1+\mathrm{i}\tan\theta} = \frac{\cos\theta - \mathrm{i}\sin\theta}{\cos\theta + \mathrm{i}\sin\theta} = \exp(-2\mathrm{i}\theta). \tag{7}$$

The reflected wave is represented by

$$R\exp(-\mathrm{i}kx) = \exp\{-\mathrm{i}(kx + 2\theta)\}. \tag{8}$$

(b) In region (1) the wave function is

$$\begin{aligned} u_1 &= \exp(\mathrm{i}kx) + R\exp(-\mathrm{i}kx) \\ &= \exp(-\mathrm{i}\theta)[\exp\{\mathrm{i}(kx+\theta)\} + \exp\{-\mathrm{i}(kx+\theta)\}] \\ &= 2\exp(-\mathrm{i}\theta)\cos(kx+\theta). \end{aligned} \tag{9}$$

The amplitude of the function at the point x is $2\cos(kx + \theta)$. In region (2) the wave function is

$$\begin{aligned} u_2 &= T\exp(-\gamma x) = (1 + R)\exp(-\gamma x) \\ &= \{1 + \exp(-2\mathrm{i}\theta)\}\exp(-\gamma x) \\ &= 2\exp(-\mathrm{i}\theta)\cos\theta\exp(-\gamma x). \end{aligned} \tag{10}$$

The amplitude at x is $2\cos\theta\exp(-\gamma x)$.

(c) Since the density of the incident particles is unity, their flux F is equal to their velocity, that is, $(2E/m)^{1/2}$. The reflected particles also have unit density, since $|R| = 1$. Their velocity is equal and opposite to that of the incident particles. Hence their flux is $-(2E/m)^{1/2}$. Inserting u_2 in the expression for F in (2.5.1) shows that the flux of particles in region (2) is zero. If u_1 is inserted in the expression for F, the result is also zero, because F is the *net* flux, and the sum of the incident and reflected flux is zero.

3.6 The Schrödinger equation has the form

$$\begin{aligned} &u'' + k^2u = 0 \quad \text{regions (1) and (3)}, \\ &u'' - \gamma^2u = 0 \quad \text{region (2)} \end{aligned} \tag{1}$$

– see Fig. 3.4. The required solutions have the form

$$\begin{aligned} u_1 &= \exp(\mathrm{i}kx) + R\exp(-\mathrm{i}kx), \\ u_2 &= A\exp(\gamma x) + B\exp(-\gamma x), \\ u_3 &= T\exp(\mathrm{i}kx). \end{aligned} \tag{2}$$

As in the previous two problems we have given the incident wave unit amplitude, and inserted the physical result that in region (3) there is no wave corresponding to particles travelling to the left.

u and u' are continuous at $x = 0$. Therefore

$$1 + R = A + B, \tag{3}$$

$$1 - R = -\mathrm{i}\rho(A - B), \quad \rho = \frac{\gamma}{k}. \tag{4}$$

u and u' are continuous at $x = a$. Therefore

$$A\exp(\theta) + B\exp(-\theta) = T\exp(\mathrm{i}\delta), \tag{5}$$

$$A\exp(\theta) - B\exp(-\theta) = \mathrm{i}\frac{T}{\rho}\exp(\mathrm{i}\delta), \tag{6}$$

where

$$\theta = \gamma a, \quad \delta = ka. \tag{7}$$

This essentially solves the problem. The rest is simply the manipulation of (3) to (6) to eliminate the coefficients R, A, B, and obtain an expression for the transmission coefficient T.

By first adding and then subtracting (5) and (6) we obtain

$$A = \tfrac{1}{2}T \exp(\mathrm{i}\delta - \theta)\left(1 + \frac{\mathrm{i}}{\rho}\right), \tag{8}$$

$$B = \tfrac{1}{2}T \exp(\mathrm{i}\delta + \theta)\left(1 - \frac{\mathrm{i}}{\rho}\right). \tag{9}$$

Add (3) and (4) to eliminate R, and replace A and B in the resulting equation by the expressions in (8) and (9). This gives

$$T = \exp(-\mathrm{i}\delta)/P,$$

where

$$P = \tfrac{1}{4}\left\{\left(1 + \frac{\mathrm{i}}{\rho}\right)\left(1 + \frac{\rho}{\mathrm{i}}\right)\exp(-\theta) + \left(1 - \frac{\mathrm{i}}{\rho}\right)\left(1 - \frac{\rho}{\mathrm{i}}\right)\exp(\theta)\right\} \tag{10}$$

$$= \cosh\theta + \tfrac{1}{2}\mathrm{i}\left(\rho - \frac{1}{\rho}\right)\sinh\theta. \tag{11}$$

Thus

$$\mathcal{T} = |T|^2 = \frac{1}{|P|^2}, \tag{12}$$

where

$$|P|^2 = 1 + \frac{(1+\rho^2)^2}{4\rho^2}\sinh^2\theta. \tag{13}$$

In (11) and (13) we have used the relations

$$\cosh\theta = \tfrac{1}{2}\{\exp(\theta) + \exp(-\theta)\},$$

$$\sinh\theta = \tfrac{1}{2}\{\exp(\theta) - \exp(-\theta)\},$$

$$\cosh^2\theta = 1 + \sinh^2\theta. \tag{14}$$

From the definition of ρ we have

$$\rho^2 = \frac{\gamma^2}{k^2} = \frac{W - E}{E}, \tag{15}$$

whence

$$\frac{(1+\rho^2)^2}{4\rho^2} = \frac{W^2}{4E(W-E)}. \tag{16}$$

Eqs. (12), (13), and (16) give the required result.

Comments

(1) Since $\mathcal{T}$ is not zero, a finite fraction of the particles pass from region (1) into region (3). They are said to *tunnel* through the potential barrier. The effect has no classical counterpart. Classically, if $W > E$, a particle incident on the barrier can only be reflected. In quantum mechanics the particles can tunnel through the barrier. Their kinetic energy in region (2) is negative, so in order to observe them in this region, we would have to disturb the system in such a way that the kinetic energy becomes positive, but that would correspond to a different potential.

The tunnelling effect is observed in several phenomena, though the shape of the potential is usually more complicated than the simple one of the present problem. α-particles emerge from an atomic nucleus with energies less than those needed to surmount the potential barrier due to the combined nuclear and electrostatic forces. In the Josephson effect, electron pairs from a superconductor tunnel through an insulating layer into another superconductor. Another example is the phenomenon known as *field emission*. When a large electric field is applied to the surface of a metal, it changes the potential at the surface, and some of the free electrons in the metal tunnel through the resulting barrier and are emitted.

(2) If $\gamma a \gg 1$, $\sinh \gamma a \gg 1$, and $\mathcal{T} \ll 1$. So the tunnelling effect is small unless $\gamma a \lesssim 1$. Suppose $(2mW)^{1/2}a/\hbar \gg 1$. Then $\mathcal{T}$ will be very small unless E is very nearly equal to W, in which case a small change in E produces a large change in the value of $\mathcal{T}$. This is shown vividly in α-particle decay, where the value of $\mathcal{T}$ is related to the lifetime of the nucleus – the larger the value of $\mathcal{T}$, i.e. the higher the probability of tunnelling through the barrier, the shorter the lifetime. For all known α-particle emitters, the value of E varies from about 2 to 8 MeV, a factor of only 4. Whereas the range of half-lives is from about 10^{11} years down to about 10^{-6} second, a factor of 10^{24}.

(3) On physical grounds, we expect F, the flux of particles, to be the same in all three regions, otherwise particles would be created or destroyed. The expression (2.5.1) for F shows that F is continuous at the two boundaries, because it involves only u and u', and we have made these two functions continuous at the boundaries. In fact the expression

for **F** in Problem 2.5 is *derived* on the assumption that particles are not created or destroyed.

Intuitively we might have thought that F_2, the flux in region (2), was zero, since the wave function u_2 does not appear to represent a travelling wave. If we insert the expression for u_2 in (2) into (2.5.1) we obtain

$$F_2 = \frac{2\hbar\gamma}{m}\,\mathrm{Im}\,(AB^*). \tag{17}$$

This shows that, for a function of the form of u_2, if the coefficient A or B is zero, or if they are both real (in general if, being complex, they have the same phase), then the flux is indeed zero. In Problem 3.5, where we have a potential barrier of infinite length, the coefficient A is zero, and so is the flux. However in the present problem neither A nor B is zero, and their relative phase gives a non-zero flux.

3.7 One method is to use the results of the last problem. If $W \to \infty$, and $a \to 0$, with $Wa = b$, then

$$\gamma^2 a^2 = \frac{2m(W-E)}{\hbar^2}a^2 \to \frac{2mW}{\hbar^2}a^2 = \frac{2mb^2}{\hbar^2 W} \to 0. \tag{1}$$

So

$$\sinh^2 \gamma a \to \gamma^2 a^2 \to \frac{2mb^2}{\hbar^2 W}. \tag{2}$$

Also

$$\frac{W^2}{E(W-E)} \to \frac{W}{E}. \tag{3}$$

From (3.6.13) to (3.6.16)

$$\frac{1}{\mathcal{T}} = |P|^2 = 1 + \frac{W^2}{4E(W-E)}\sinh^2 \gamma a \to 1 + \frac{mb^2}{2\hbar^2 E}, \tag{4}$$

which is the required result.

Alternatively, we can obtain the result for a δ-function potential by a direct method. The Schrödinger equation for region (2) of the last problem is

$$u'' - \gamma^2 u = 0. \tag{5}$$

As $W \to \infty$

$$\gamma^2 = \frac{2m(W-E)}{\hbar^2} \to \frac{2mW}{\hbar^2} \to \infty. \tag{6}$$

Since u is finite in region (2), eqs. (5) and (6) show that u'' tends to infinity in that region. The value of u' increases between $x = 0$ and $x = a$ by approximately $u''a$, a finite quantity, i.e.

$$u'(a) - u'(0) \approx u''a \approx \gamma^2 au, \tag{7}$$

where

$$\gamma^2 a \approx \frac{2mWa}{\hbar^2} \approx \frac{2m}{\hbar^2} b. \tag{8}$$

Since u' is finite in region (2), the change in u between $x = 0$ and $x = a$ tends to zero as $a \to 0$, i.e.

$$u(a) \approx u(0). \tag{9}$$

Applying the results of (7) and (9) to the functions u_1, u_2, u_3 in (3.6.2) we obtain

$$1 + R = T, \tag{10}$$

$$ikT - ik(1 - R) = \frac{2m}{\hbar^2} bT. \tag{11}$$

Eliminating R between these two equations gives

$$\frac{1}{T} = 1 + \frac{imb}{\hbar^2 k}. \tag{12}$$

Thus

$$\frac{1}{\mathcal{T}} = \frac{1}{|T|^2} = 1 + \frac{m^2 b^2}{\hbar^4 k^2} = 1 + \frac{mb^2}{2\hbar^2 E}. \tag{13}$$

4

Orbital angular momentum, hydrogen atom, harmonic oscillator

Summary of theory

1 Orbital angular momentum

(a) Operators, eigenfunctions, eigenvalues

L_x, L_y, L_z are operators corresponding to x, y, z components of orbital angular momentum. The operator corresponding to the square of the magnitude of the orbital angular momentum is

$$L^2 = L_x^2 + L_y^2 + L_z^2. \tag{1}$$

L^2 commutes with L_x, L_y, and L_z. The component operators do not commute with each other. The commutation relations are

$$[L_x, L_y] = \mathrm{i}\hbar L_z, \quad \text{and cyclic permutations.} \tag{2}$$

The common eigenfunctions of L^2 and L_z are the spherical harmonics Y_{lm}. The eigenvalues of L^2 and L_z are given by

$$L^2 Y_{lm} = l(l+1)\hbar^2 Y_{lm}, \quad L_z Y_{lm} = m\hbar Y_{lm}, \tag{3}$$

where l is zero, or a positive integer. For given l, m takes the values

$$l, l-1, l-2, \ldots 1, 0, -1, \ldots -l. \tag{4}$$

The spherical harmonics Y_{lm} are functions of the polar angles θ, ϕ. They are orthogonal, and are conventionally normalised so that the square of the modulus of the function integrated over all directions is unity, i.e.

$$\begin{aligned}\int_0^{\pi}\int_0^{2\pi} Y^*_{l'm'} Y_{lm} \sin\theta \,\mathrm{d}\theta\,\mathrm{d}\phi &= 0, \quad \text{unless } l' = l, \text{ and } m' = m,\\ &= 1, \quad l' = l,\ m' = m.\end{aligned} \tag{5}$$

The functions for the values $l = 0, 1, 2$ are listed in Table 4.1.

Table 4.1. *Spherical harmonics for $l = 0, 1, 2$.*

$$Y_{00} = \frac{1}{(4\pi)^{1/2}}$$

$$Y_{10} = \left(\frac{3}{4\pi}\right)^{1/2} \cos\theta$$

$$Y_{1\pm1} = \mp\left(\frac{3}{8\pi}\right)^{1/2} \sin\theta \exp(\pm i\phi)$$

$$Y_{20} = \left(\frac{5}{16\pi}\right)^{1/2} (3\cos^2\theta - 1)$$

$$Y_{2\pm1} = \mp\left(\frac{15}{8\pi}\right)^{1/2} \cos\theta \sin\theta \exp(\pm i\phi)$$

$$Y_{2\pm2} = \left(\frac{15}{32\pi}\right)^{1/2} \sin^2\theta \exp(\pm 2i\phi)$$

(b) Ladder operators

The operators L_+, L_- are defined by

$$L_+ = L_x + iL_y, \quad L_- = L_x - iL_y. \tag{6}$$

They satisfy the relations

$$L_+ Y_{lm} = \{l(l+1) - m(m+1)\}^{1/2} \hbar Y_{l,m+1}, \tag{7}$$

$$L_- Y_{lm} = \{l(l+1) - m(m-1)\}^{1/2} \hbar Y_{l,m-1}. \tag{8}$$

For a fixed value of l, the set of eigenfunctions Y_{lm}, where m takes integral values from $-l$ to $+l$, may be regarded as arranged on a ladder with $2l + 1$ rungs. The operator L_+ converts the function Y_{lm} into the function one rung *up* the ladder, while the operator L_- converts it into the function one rung *down*; hence the name *ladder operators* for L_+ and L_-. The operator L_+ is also called a *raising operator*, and L_- a *lowering operator*. From (7) and (8)

$$L_+ Y_{ll} = L_- Y_{l,-l} = 0, \tag{9}$$

which is consistent with the fact that the ladder is bounded at $m = \pm l$.

2 Hydrogen-type atom

Consider an atom or ion with a single electron and nuclear charge Ze; ignore the spin of the electron. $Z = 1$ refers to a hydrogen atom, $Z = 2$

to a He^+ ion, $Z = 3$ to a Li^{++} ion, and so on. Assume the mass of the nucleus is infinitely large compared to the mass of the electron, so that the nucleus is at rest, and use spherical polar coordinates r, θ, ϕ, with the origin at the position of the nucleus. The potential of the electron, due to the electrostatic force of the nucleus, is

$$V(r) = -\frac{Ze^2}{4\pi\varepsilon_0}\frac{1}{r}. \tag{10}$$

The eigenfunctions of the Hamiltonian are specified by the three quantum numbers n, l, m and may be expressed as

$$u_{nlm}(r, \theta, \phi) = R_{nl}(r)Y_{lm}(\theta, \phi), \tag{11}$$

where Y_{lm} is a spherical harmonic. The quantum number n is a positive integer, and, for a fixed value of n, l takes integral values from zero to $n - 1$. The energy of the state u_{nlm} depends only on n, and not on the other two quantum numbers. It is given by

$$E_n = -\frac{m_e}{2\hbar^2}\left(\frac{Ze^2}{4\pi\varepsilon_0}\right)^2\frac{1}{n^2}. \tag{12}$$

The radial function $R_{nl}(r)$ has the form

$$R_{nl}(r) = c_{nl}\exp\left(-\frac{Zr}{na_0}\right)W_{nl}(r), \tag{13}$$

where

$$a_0 = \frac{\hbar^2}{m_e}\frac{4\pi\varepsilon_0}{e^2}. \tag{14}$$

a_0 has the dimensions of length, and is known as the *Bohr radius*. The function $W_{nl}(r)$ is a polynomial in r, with powers of r going from l up to $n - 1$. The normalising constant c_{nl} is defined so that

$$\int_0^\infty \{R_{nl}(r)\}^2 r^2\,\mathrm{d}r = 1. \tag{15}$$

The expressions for $R_{nl}(r)$ for $n = 1$ and 2 are given in Table 4.2. When the finite mass of the nucleus is taken into account, the expressions for the energy and R_{nl} remain the same, except that m_e, the mass of the electron, is replaced by the reduced mass of the electron–nucleus system.

It may be noted that if the potential $V(r)$ is a function only of the magnitude of $\mathbf{r}$ (representing a *central* force field), the angular part of the solution of the Schrödinger equation is a spherical harmonic $Y_{lm}(\theta, \phi)$. This is true whatever the functional form of $V(r)$. The function $V(r)$ determines the radial part of the solution.

Table 4.2. *The radial function R_{nl} for n = 1, 2.*

$$R_{10} = 2\left(\frac{Z}{a_0}\right)^{3/2} \exp\left(-\frac{Zr}{a_0}\right)$$
$$R_{20} = \frac{1}{\sqrt{2}}\left(\frac{Z}{a_0}\right)^{3/2}\left(1 - \frac{Zr}{2a_0}\right) \exp\left(-\frac{Zr}{2a_0}\right)$$
$$R_{21} = \frac{1}{\sqrt{24}}\left(\frac{Z}{a_0}\right)^{5/2} r \exp\left(-\frac{Zr}{2a_0}\right)$$

3 Harmonic oscillator

For a one-dimensional oscillator of mass m and angular frequency ω, the Hamiltonian can be expressed as

$$H = \frac{1}{2m}(m^2\omega^2x^2 + p^2), \tag{16}$$

where x is the displacement, and p is the momentum. We define operators a and $a^\dagger$ by

$$a = (2m\hbar\omega)^{-1/2}(m\omega x + ip),$$
$$a^\dagger = (2m\hbar\omega)^{-1/2}(m\omega x - ip). \tag{17}$$

From (16) and (17), and the commutation relation $[p, x] = -i\hbar$, the following results may be deduced.

(i) $$aa^\dagger = \frac{H}{\hbar\omega} + \tfrac{1}{2}, \quad a^\dagger a = \frac{H}{\hbar\omega} - \tfrac{1}{2}, \tag{18}$$

$$[a, a^\dagger] = 1. \tag{19}$$

(ii) The eigenvalues of the Hamiltonian are

$$E_n = (n + \tfrac{1}{2})\hbar\omega, \tag{20}$$

where n is zero or a positive integer.

(iii) If u_n is a normalised eigenfunction of the Hamiltonian corresponding to the eigenvalue E_n, then

$$au_n = n^{1/2}u_{n-1}, \quad a^\dagger u_n = (n + 1)^{1/2}u_{n+1}. \tag{21}$$

These relations show that a and $a^\dagger$ act as ladder operators. The operator a is known as an *annihilation operator*, and $a^\dagger$ as a *creation operator*.

Problems

4.1 A system is in the state $\psi = \phi_{lm}$, an eigenstate of the angular momentum operators L^2 and L_z. Calculate $\langle L_x \rangle$, and $\langle L_x^2 \rangle$.

4.2 Denote the eigenfunctions of the orbital angular momentum operators L^2 and L_z with eigenvalues $l = 1$ and $m = 1, 0, -1$ by $\phi_1, \phi_0, \phi_{-1}$. Use the raising and lowering operators L_+ and L_- to calculate the result of operating on $\phi_1, \phi_0, \phi_{-1}$ with L_x. Hence find the eigenfunctions and eigenvalues of L_x.

4.3 (a) A beam of particles with $l = 1$ is travelling along the y axis and passes through a Stern–Gerlach magnet P with its (mean) magnetic field along the x axis – Fig. 4.1. The emerging beam with $m = 1$ is separated from the other two and passed through a second Stern–Gerlach magnet Q with the magnetic field along the z axis. Into how many beams is this beam further split, and what are the relative numbers of atoms in them?

(b) What would be the results if the $m = 0$ and $m = -1$ beams from magnet P were treated in the same way?

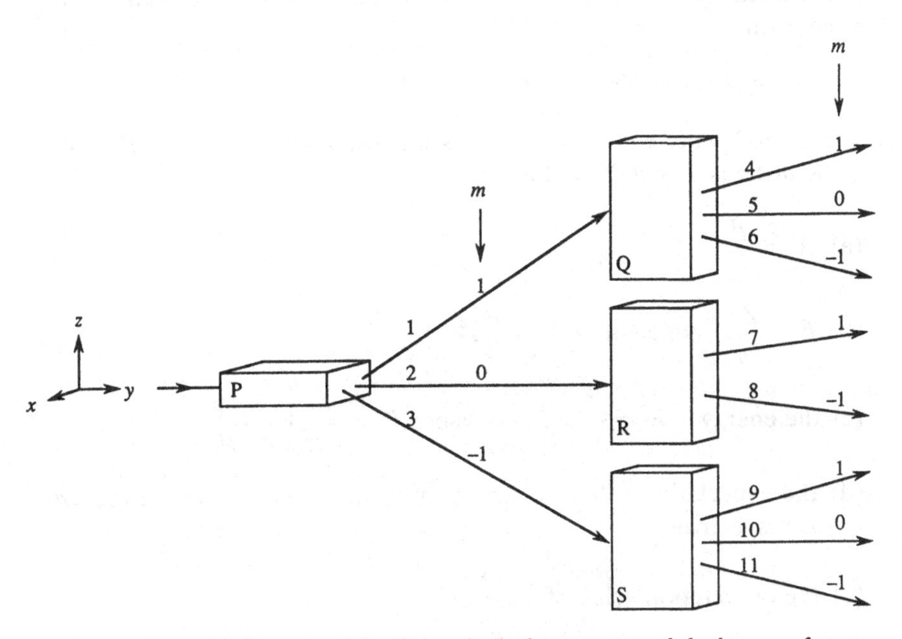

Fig. 4.1. *Schematic diagram of the Stern–Gerlach magnets and the beams of atoms for Problems 4.3 and 4.4. The m values for beams 1–3 refer to the x axis, and the m values for beams 4–11 refer to the z axis.*

4.4 Consider the following description of the measurements in Problem 4.3. Magnet P in Fig. 4.1 measures L_x, the x-component of angular momentum, and the atoms in beam 1 have $L_x = \hbar$. Magnet Q then measures the z-component of angular momentum of the atoms in this beam. The atoms in beam 4 have $L_x = \hbar$, and $L_z = \hbar$. Therefore the two magnets together have measured the x- and z-components of angular momentum.

(a) Is this correct? If not, give the correct description.

(b) What would happen if beam 4 were passed into another Stern–Gerlach magnet with its magnetic field along x?

4.5 α and β are physical quantities with corresponding operators A and B. Are the following statements true or false?

(a) If A and B commute and we know α with certainty, we know β with certainty.

(b) If A and B do not commute, we cannot know both α and β with certainty.

If false give counter examples.

4.6 The normalised wave function for the ground state of a hydrogen-like atom (neutral hydrogen, He^+, Li^{++}, etc.) with nuclear charge Ze, has the form

$$u = A \exp(-\beta r),$$

where A and β are constants, and r is the distance between the electron and the nucleus. Show the following.

(a) $A^2 = \dfrac{\beta^3}{\pi}$,

(b) $\beta = \dfrac{Z}{a_0}$, where $a_0 = \dfrac{\hbar^2}{m_e}\dfrac{4\pi\varepsilon_0}{e^2}$,

(c) the energy is $E = -Z^2 E_0$, where $E_0 = \dfrac{m_e}{2\hbar^2}\left(\dfrac{e^2}{4\pi\varepsilon_0}\right)^2$,

(d) the expectation values of the potential and kinetic energies are $2E$ and $-E$ respectively,

(e) the expectation value of r is $\dfrac{3a_0}{2Z}$,

(f) the most probable value of r is $\dfrac{a_0}{Z}$.

[The operator ∇^2 becomes $\partial^2/\partial r^2 + (2/r)\,\partial/\partial r$ when operating on a function that depends only on the magnitude of r.]

4.7 An atom of tritium is in its ground state, when the nucleus suddenly decays into a helium nucleus, with the emission of a fast electron which leaves the atom without perturbing the extranuclear electron. Find the probability that the resulting He^+ ion will be left in

(a) a 1*s* state,

(b) a 2*s* state.

(c) What is the selection rule for the l quantum number in the transition?

4.8 Explain how the solutions of the Schrödinger equation for the hydrogen atom are related to a z axis despite the fact that the potential is spherically symmetric.

4.9 Prove the following results for an eigenstate of the Hamiltonian of a one-dimensional harmonic oscillator.

(a) The expectation values of the position and momentum are zero,

(b) The expectation values of the potential and kinetic energies are equal.

(c) Δ_x and Δ_p, the uncertainties in position and momentum, satisfy the relation $\Delta_x\Delta_p = (n+\frac{1}{2})\hbar$, where n is the quantum number of the state.

(d) Show that the ground-state eigenfunction u_0 is a Gaussian, and hence that the parity of the state n is even for even n, and odd for odd n.

4.10 The Hamiltonian of a three-dimensional isotropic harmonic oscillator with mass m and angular frequency ω has the form

$$H = -\frac{\hbar^2}{2m}\left(\frac{\partial^2}{\partial x^2} + \frac{\partial^2}{\partial y^2} + \frac{\partial^2}{\partial z^2}\right) + \tfrac{1}{2}m\omega^2(x^2 + y^2 + z^2).$$

(a) By expressing H as the sum of three similar Hamiltonians H_x, H_y, H_z for one-dimensional oscillators, show that the energy is equal to $(n+\frac{3}{2})\hbar\omega$, where n is zero or a positive integer.

(b) Show that the degeneracies of the three lowest energy values are 1, 3, 6, and that in general the degeneracy of the level n is $\frac{1}{2}(n+1)(n+2)$.

(c) Deduce the parity of the states of the level n.

Solutions

4.1 From (4.6), (4.7), and (4.8) we have the following results.

$$L_x = \tfrac{1}{2}(L_+ + L_-), \tag{1}$$

$$L_+\phi_{lm} = c_+\phi_{l,m+1}, \tag{2}$$

$$L_-\phi_{lm} = c_-\phi_{l,m-1}, \tag{3}$$

$$L_+L_- + L_-L_+ = 2(L^2 - L_z^2). \tag{4}$$

It is only the form of (2) and (3) that we require in the present problem. The values of the constants c_+ and c_- are irrelevant.

The formal solution is as follows.

$$\langle L_x \rangle = \int \phi_{lm}^* L_x \phi_{lm}\,\mathrm{d}\tau = \tfrac{1}{2}\int \phi_{lm}^*(L_+ + L_-)\phi_{lm}\,\mathrm{d}\tau. \tag{5}$$

Now

$$L_+\phi_{lm} = c_+\phi_{l,m+1}, \tag{6}$$

and

$$\int \phi_{lm}^*\phi_{l,m+1}\,\mathrm{d}\tau = 0, \tag{7}$$

since ϕ_{lm} and $\phi_{l,m+1}$ are orthogonal. Therefore

$$\int \phi_{lm} L_+ \phi_{lm}\,\mathrm{d}\tau = 0. \tag{8}$$

Similarly

$$\int \phi_{lm}^* L_- \phi_{lm}\,\mathrm{d}\tau = 0. \tag{9}$$

Thus

$$\langle L_x \rangle = 0. \tag{10}$$

The expectation value of L_x^2 is

$$\langle L_x^2 \rangle = \int \phi_{lm}^* L_x^2 \phi_{lm}\,\mathrm{d}\tau, \tag{11}$$

where

$$L_x^2 = \tfrac{1}{4}(L_+ + L_-)^2 = \tfrac{1}{4}(L_+L_- + L_-L_+ + L_+^2 + L_-^2). \tag{12}$$

As before

$$\int \phi_{lm}^* L_+^2 \phi_{lm}\,\mathrm{d}\tau = \text{constant} \times \int \phi_{lm}^* \phi_{l,m+2}\,\mathrm{d}\tau = 0. \tag{13}$$

Similarly

$$\int \phi_{lm}^* L_-^2 \phi_{lm} \, d\tau = 0. \tag{14}$$

Therefore

$$\begin{aligned} \langle L_x^2 \rangle &= \tfrac{1}{4}\int \phi_{lm}^*(L_+L_- + L_-L_+)\phi_{lm} \, d\tau \\ &= \tfrac{1}{2}\int \phi_{lm}^*(L^2 - L_z^2)\phi_{lm} \, d\tau \\ &= \tfrac{1}{2}\{l(l+1) - m^2\}\hbar^2, \end{aligned} \tag{15}$$

since

$$(L^2 - L_z^2)\phi_{lm} = \{l(l+1) - m^2\}\hbar^2 \phi_{lm}, \tag{16}$$

and ϕ_{lm} is normalised.

These two results may be obtained much more quickly by physical reasoning. The state ϕ_{lm} corresponds to one in which the magnitude of the angular momentum and its z-component are known. However, the x axis is not defined and may be anywhere in the plane perpendicular to the z axis. Therefore $\langle L_x \rangle$ must be zero, otherwise different observers choosing different x axes would get different values for $\langle L_x \rangle$.

By the same reasoning

$$\langle L_x^2 \rangle = \langle L_y^2 \rangle, \quad \text{but} \quad \langle L_x^2 \rangle + \langle L_y^2 \rangle + \langle L_z^2 \rangle = \langle L^2 \rangle. \tag{17}$$

Therefore

$$\langle L_x^2 \rangle = \tfrac{1}{2}\{\langle L^2 \rangle - \langle L_z^2 \rangle\} = \tfrac{1}{2}\{l(l+1) - m^2\}\hbar^2. \tag{18}$$

4.2 Put $l = 1$ in (4.7) and (4.8). Then

$$\begin{aligned} L_+\phi_1 &= 0, & L_-\phi_1 &= \sqrt{2}\hbar\phi_0, \\ L_+\phi_0 &= \sqrt{2}\hbar\phi_1, & L_-\phi_0 &= \sqrt{2}\hbar\phi_{-1}, \\ L_+\phi_{-1} &= \sqrt{2}\hbar\phi_0, & L_-\phi_{-1} &= 0. \end{aligned} \tag{1}$$

To find the eigenfunctions of L_x in terms of ϕ_1, ϕ_0, and ϕ_{-1}, we first use (1) to obtain the result of L_x operating on these functions. Since $L_x = \tfrac{1}{2}(L_+ + L_-)$, we have

$$L_x\phi_1 = \frac{\hbar}{\sqrt{2}}\phi_0, \quad L_x\phi_0 = \frac{\hbar}{\sqrt{2}}(\phi_1 + \phi_{-1}), \quad L_x\phi_{-1} = \frac{\hbar}{\sqrt{2}}\phi_0. \tag{2}$$

If $p\phi_1 + q\phi_0 + r\phi_{-1}$, where p, q, r are constants to be determined, is an eigenfunction of L_x with eigenvalue λ, then

$$L_x(p\phi_1 + q\phi_0 + r\phi_{-1}) = \lambda(p\phi_1 + q\phi_0 + r\phi_{-1}). \tag{3}$$

Insert the results of (2) in the left-hand side of (3), and equate the coefficients of ϕ_1, ϕ_0, and ϕ_{-1} on the two sides of the equation. This gives

$$\frac{\hbar}{\sqrt{2}}q = \lambda p, \quad \frac{\hbar}{\sqrt{2}}(p + r) = \lambda q, \quad \frac{\hbar}{\sqrt{2}}q = \lambda r. \tag{4}$$

These equations have three solutions:

$$\lambda = 0, \qquad \text{with } q = 0,\ r = -p, \tag{5}$$

$$\lambda = \hbar, \qquad \text{with } q = \sqrt{2}p,\ r = p, \tag{6}$$

$$\lambda = -\hbar, \quad \text{with } q = -\sqrt{2}p,\ r = p. \tag{7}$$

The eigenfunctions are normalised by putting $p^2 + q^2 + r^2 = 1$. This gives the following results for the eigenfunctions and eigenvalues of L_x.

Eigenfunction	Eigenvalue
$(\phi_1 + \sqrt{2}\phi_0 + \phi_{-1})/2$	$\hbar$
$(\phi_1 - \phi_{-1})/\sqrt{2}$	0
$(\phi_1 - \sqrt{2}\phi_0 + \phi_{-1})/2$	$-\hbar$

(8)

Comment

The eigenvalues for L_x are the same as those of L_z, as of course they must be. There is nothing special about the z axis, or any other direction in space. We could have done the calculation, i.e. finding the eigenfunctions and eigenvalues, for any specified direction in space. The eigenfunctions would depend on the direction, but the eigenvalues would be $\hbar$, 0, $-\hbar$ for all directions.

4.3 (a) The state of the beam with $m = 1$ emerging from the x-magnet P is the eigenfunction of the operator L_x with eigenvalue $\hbar$. (The term x-magnet refers to a Stern–Gerlach magnet with its mean magnetic field along the x axis.) Denote this state by ρ_1. Passing a single atom of the beam through the z-magnet Q is equivalent to measuring its z-component of angular momentum. The only possible results of the measurement are the eigenvalues of the operator L_z, namely, $\hbar$, 0, $-\hbar$. The probability of obtaining, say, the value $\hbar$, which is equal to the fractional number of atoms in the beam emerging from the z-magnet with $m = 1$, is given by expanding ρ_1 in terms of ϕ_1, ϕ_0, ϕ_{-1}, the eigenfunctions of L_z, and taking the square of the modulus of the coefficient of ϕ_1. From (4.2.8)

$$\rho_1 = (\phi_1 + \sqrt{2}\phi_0 + \phi_{-1})/2. \tag{1}$$

The fractional number of atoms in the output beam of the z-magnet with $L_z = \hbar$ is therefore $(\frac{1}{2})^2 = \frac{1}{4}$. The fractional numbers of atoms in the output beams with $L_z = 0$, $L_z = -\hbar$ is obtained in a similar manner. So the answer is that the $m = 1$ beam emerging from magnet P is split into three by magnet Q, with fractional numbers of atoms equal to $\frac{1}{4}, \frac{1}{2}, \frac{1}{4}$.

(b) The atoms emerging from magnet P with $m = 0$, are in the state

$$\rho_0 = (\phi_1 - \phi_{-1})/\sqrt{2} \tag{2}$$

(from 4.2.8). So if this beam is passed through the z-magnet R, the emerging atoms have either $L_z = \hbar$ or $L_z = -\hbar$, with fractional numbers given by the square of the coefficients of ϕ_1 and ϕ_{-1}, i.e. $\frac{1}{2}, \frac{1}{2}$.

The atoms from magnet P with $m = -1$, are in the state

$$\rho_{-1} = (\phi_1 - \sqrt{2}\phi_0 + \phi_{-1})/2. \tag{3}$$

The fractional number in the beams emerging from the z-magnet S for this input beam is $\frac{1}{4}, \frac{1}{2}, \frac{1}{4}$.

Note that for each of the three beams emerging from the x-magnet P, the fractional numbers with $L_z = \hbar$, and $L_z = -\hbar$ in the outputs from the z-magnets Q, R, S are equal. This must be so, because, as we saw in Problem 4.1, for a system in an eigenstate of L^2 and L_x, the expectation value of L_z is zero.

4.4 (a) The last two sentences in the description are incorrect. It is correct that the atoms in beam 1 in Fig. 4.1 have x-component $L_x = \hbar$, but this is not so for the atoms in beam 4. Magnet Q measures the z-component of angular momentum, and the atoms in beam 4 have $L_z = \hbar$, but the measurement of L_z disturbs the atoms so that the information concerning the x-component of angular momentum is lost.

(b) The state of the atoms in beam 4 is given by the function ϕ_1. To find out what will happen if beam 4 is passed through another x-magnet, we must express ϕ_1 in terms of $\rho_1, \rho_0, \rho_{-1}$, the eigenfunctions of L_x. Using the reciprocal form of (4.2.8) we have

$$\phi_1 = (\rho_1 + \sqrt{2}\rho_0 + \rho_{-1})/2. \tag{1}$$

So three beams will emerge from the second x-magnet with fractional numbers $\frac{1}{4}, \frac{1}{2}, \frac{1}{4}$. If we *had* known the x-component of angular momentum for beam 4, only *one* beam would have emerged.

Note that beams 4, 7, and 9 in Fig. 4.1 are all in the identical state ϕ_1, and all would give the same set of fractional numbers if they were passed through another x-magnet, or any other measuring apparatus. The fact that they came from different input beams (1, 2, and 3) is irrelevant.

4.5 Both statements are in general false.
(a) If the operators A and B commute we can find a set of functions which are eigenfunctions of both operators. However, if some eigenvalues are degenerate for one or both operators, it is possible to have functions which are eigenfunctions of one operator but not of the other. Consider for example the functions ϕ_{lm}, which are eigenfunctions of L^2, the operator for the square of the magnitude of angular momentum, and L_z, the operator corresponding to the z-component of angular momentum. (The discussion refers to orbital angular momentum.) L^2 and L_z commute. The functions ϕ_{lm} are not eigenfunctions of L_x and L_y (except for the special case ϕ_{00}), despite the fact that both L_x and L_y commute with L^2. Therefore, for a system in a state $\psi = \phi_{lm}$ we know the value of L^2 with certainty, but not the value of L_x (or L_y). That is to say, if we measure the value of L_x for a set of systems all in the same state ϕ_{lm}, we shall get a variety of values.

For a fixed l, there are $2l + 1$ ϕ_{lm} functions that are degenerate for the operator L^2. We may therefore take linear combinations of these functions, which remain eigenfunctions of L^2 with the same eigenvalue. By suitable choice of the coefficients we can construct combinations that are eigenfunctions of L_x, but they will not be eigenfunctions of L_z. We have seen this in detail for $l = 1$ in the last three problems.

(b) If A and B commute, it means that

$$AB\psi = BA\psi \tag{1}$$

for *any* function ψ. However, if A and B do not commute, although the above relation is not true in general, it may be the case that

$$AB\psi_0 = BA\psi_0 \tag{2}$$

for some special function ψ_0. An example of this is the function ϕ_{00}, which is an eigenfunction of the three non-commuting operatore L_x, L_y, L_z. So for a system in this state we known the three components of angular momentum with certainty. They are all zero.

4.6 We shall need the following results.

$$\text{(i)} \quad \int_0^\infty r \exp(-\mu r)\,\mathrm{d}r = \frac{1}{\mu^2}, \quad \int_0^\infty r^2 \exp(-\mu r)\,\mathrm{d}r = \frac{2}{\mu^3},$$

$$\int_0^\infty r^3 \exp(-\mu r)\,\mathrm{d}r = \frac{6}{\mu^4}. \tag{1}$$

They may be verified by differentiation by parts. Alternatively, the last two may be obtained from the first by differentiating both sides with respect to μ.

(ii) For a symmetrically symmetric function $f(r)$

$$\int_{\substack{\text{all}\\\text{space}}} f(r)\,\mathrm{d}\tau = 4\pi\int_0^\infty r^2 f(r)\,\mathrm{d}r. \tag{2}$$

(a) The constant A in the wave function is given by the normalisation condition

$$\int_{\substack{\text{all}\\\text{space}}} |u|^2\,\mathrm{d}\tau = 1. \tag{3}$$

From (1) and (2)

$$\int_{\substack{\text{all}\\\text{space}}} |u|^2\,\mathrm{d}\tau = 4\pi A^2\int_0^\infty r^2\exp(-2\beta r)\,\mathrm{d}r = \frac{\pi A^2}{\beta^3} = 1. \tag{4}$$

Thus

$$A^2 = \frac{\beta^3}{\pi}. \tag{5}$$

(b) The Schrödinger equation is

$$-\frac{\hbar^2}{2m_\mathrm{e}}\nabla^2 u + V(r)u = Eu. \tag{6}$$

Insert $u = A\exp(-\beta r)$ into (6), and use the results

$$\begin{aligned}\nabla^2\exp(-\beta r) &= \left(\frac{\partial^2}{\partial r^2} + \frac{2}{r}\frac{\partial}{\partial r}\right)\exp(-\beta r)\\ &= \left(\beta^2 - \frac{2\beta}{r}\right)\exp(-\beta r),\end{aligned} \tag{7}$$

$$V(r) = -\frac{\alpha}{r}, \quad \text{where} \quad \alpha = \frac{Ze^2}{4\pi\varepsilon_0}. \tag{8}$$

This gives

$$-\frac{\hbar^2}{2m_\mathrm{e}}\left(\beta^2 - \frac{2\beta}{r}\right) - \frac{\alpha}{r} = E. \tag{9}$$

The function $u = A\exp(-\beta r)$ is a solution of (6) if the terms in $1/r$ in (9) add to zero, and the other two terms are equal. The first condition gives

$$\beta = \frac{m_\mathrm{e}}{\hbar^2}\alpha = \frac{m_\mathrm{e}}{\hbar^2}\frac{Ze^2}{4\pi\varepsilon_0} = \frac{Z}{a_0}. \tag{10}$$

(c) The second condition gives

$$E = -\frac{\hbar^2}{2m_e}\beta^2 = -Z^2\frac{m_e}{2\hbar^2}\left(\frac{e^2}{4\pi\varepsilon_0}\right)^2 = -Z^2E_0. \tag{11}$$

(d) The expectation value of the potential energy is

$$\begin{aligned}\langle V\rangle &= \int u^*V(r)u\,\mathrm{d}\tau = -4\pi A^2\alpha\int_0^\infty r\exp(-2\beta r)\,\mathrm{d}r\\ &= -4\alpha\beta^3\frac{1}{4\beta^2} = -\alpha\beta = -\frac{\hbar^2}{m_e}\beta^2 = 2E.\end{aligned} \tag{12}$$

We could calculate the expectation value of the kinetic energy T from

$$\langle T\rangle = -\frac{\hbar^2}{2m_e}\int u^*\nabla^2u\,\mathrm{d}\tau, \tag{13}$$

but it is much quicker to use the relation

$$\langle T\rangle = \langle E\rangle - \langle V\rangle = E - 2E = -E. \tag{14}$$

(e) The expectation value of r is

$$\begin{aligned}\langle r\rangle &= \int u^*ru\,\mathrm{d}\tau = 4\pi A^2\int_0^\infty r^3\exp(-2\beta r)\,\mathrm{d}r\\ &= 4\beta^3\frac{6}{16\beta^4} = \frac{3}{2}\frac{a_0}{Z}.\end{aligned} \tag{15}$$

(f) The probability of finding the electron between r and $r + \mathrm{d}r$ from the nucleus is $P(r)\,\mathrm{d}r$, where

$$P(r) = 4\pi r^2|u(r)|^2 = 4\pi A^2r^2\exp(-2\beta r). \tag{16}$$

Differentiating with respect to r, shows that $P(r)$ is a maximum when

$$r = \frac{1}{\beta} = \frac{a_0}{Z}. \tag{17}$$

Comment

From (12) and (14) we see that

$$\langle T\rangle = -\tfrac{1}{2}\langle V\rangle. \tag{18}$$

This is a special case of a general result, known as the *virial theorem*, which says that, for a system in a stationary state in a potential $V(r)$ proportional to r^n,

$$\langle T\rangle = \frac{n}{2}\langle V\rangle. \tag{19}$$

For the present case of an inverse-square law of force, $n = -1$, which gives the result in (18). The virial theorem holds also in classical mechanics. Thus the result in (18) applies to a satellite orbiting the Earth, the force being the inverse-square law of gravitation. For proofs and discussion of the virial theorem in quantum and classical mechanics see Davydov (1965), p. 51 and Goldstein (1980), p. 82.

4.7 The wave function of the electron immediately after the nuclear decay is $(u_{1s})_{Z=1}$, i.e. the 1*s* state for the hydrogen atom. Denote the states of the He^+ ion by u_i. Then the probability that the He^+ ion is in a particular state j is equal to $|c_j|^2$, where c_j is the coefficient of u_j in the expansion

$$(u_{1s})_{Z=1} = \sum_i c_i u_i. \tag{1}$$

From (2.5) c_j is given by

$$c_j = \int_{\substack{\text{all}\\\text{space}}} u_j^*(u_{1s})_{Z=1}\,\mathrm{d}\tau. \tag{2}$$

From (4.11) and Tables 4.1 and 4.2, the relevant wave functions are

$$(u_{1s})_{Z=1} = (1/\pi a_0^3)^{1/2} \exp(-r/a_0), \tag{3}$$

$$(u_{1s})_{Z=2} = (8/\pi a_0^3)^{1/2} \exp(-2r/a_0), \tag{4}$$

$$(u_{2s})_{Z=2} = (1/\pi a_0^3)^{1/2}\left(1 - \frac{r}{a_0}\right) \exp(-r/a_0). \tag{5}$$

Since these functions depend only on r, we evaluate the integral in (2) by putting

$$\mathrm{d}\tau = 4\pi r^2\,\mathrm{d}r, \tag{6}$$

and integrating with respect to r from zero to infinity.

(a) The probability that the helium ion is in the state 1*s* is $|c_{1s}|^2$, where

$$\begin{aligned} c_{1s} &= \int_{\substack{\text{all}\\\text{space}}} (u_{1s})^*_{Z=2}(u_{1s})_{Z=1}\,\mathrm{d}\tau \\ &= 4\pi\frac{8^{1/2}}{\pi a_0^3}\int_0^\infty r^2 \exp(-3r/a_0)\,\mathrm{d}r \\ &= 2^{7/2}\int_0^\infty x^2 \exp(-3x)\,\mathrm{d}x = \frac{2^{9/2}}{3^3}. \end{aligned} \tag{7}$$

In the last step we have used one of the results in (4.6.1). The required probability is thus

$$|c_{1s}|^2 = \frac{2^9}{3^6} = 0.702. \tag{8}$$

(b) The probability that the helium ion is in the state $2s$ is $|c_{2s}|^2$, where

$$\begin{aligned} c_{2s} &= \int_{\substack{\text{all}\\\text{space}}} (u_{2s})^*_{Z=2}(u_{1s})_{Z=1}\,\mathrm{d}\tau \\ &= 4\pi\frac{1}{\pi a_0^3}\int_0^\infty r^2\left(1-\frac{r}{a_0}\right)\exp(-2r/a_0)\,\mathrm{d}r \\ &= 4\int_0^\infty x^2(1-x)\exp(-2x)\,\mathrm{d}x = -\tfrac{1}{2}. \end{aligned} \tag{9}$$

In the last step we have used two of the results in (4.6.1). The required probability is thus

$$|c_{2s}|^2 = 0.25. \tag{10}$$

(c) The selection rule is $\Delta l = 0$. This follows from the fact that the angular part of the wave function is a spherical harmonic, and the spherical harmonics for different l values are orthogonal. The function $(u_{1s})_{Z=1}$ in (1) has $l = 0$. Therefore, if the function u_j corresponds to a spherical harmonic with a value of l other than zero, the integral in (2) is zero. The selection rule reflects the fact that the angular momentum of the extranuclear electron is unchanged by the decay of the nucleus. Therefore, the value of the quantum number l, which gives the magnitude of the angular momentum of the electron, is unchanged.

4.8 The solutions of the non-relativistic Schrödinger equation for the hydrogen atom may be written as

$$u_{nlm}(r, \theta, \phi) = R_{nl}(r)Y_{lm}(\theta, \phi), \tag{1}$$

where n is the principal, l the orbital, and m the magnetic quantum number. The angular dependence of u_{nlm} lies in that of the spherical harmonics $Y_{lm}(\theta, \phi)$. These functions are not spherically symmetric, except for $l = 0$.

This may appear paradoxical, inasmuch as the potential, $-e^2/4\pi\varepsilon_0 r$, is spherically symmetric. The answer is that the potential is only spherically symmetric if there is no external force, such as that provided by an external magnetic field. A non-spherically symmetric eigenfunction u_{nlm} can only represent the state of the atom if an external force exists, or has existed at some earlier time. Such a force provides a *physically* defined z axis. In the absence of such a force, the state of the atom is represented, either by a spherically symmetric $l = 0$ state, or by a *mixture* of non-spherically symmetric states, the mixture itself being spherically symmetric. Let us see how this comes about.

Suppose we have a box of hydrogen atoms, all of which are in states with $n = 2$, $l = 1$. No axis has been defined in any physical way. We now measure the component of angular momentum along some constant axis for each of the atoms in turn. This may be done by allowing the atoms to pass through a Stern–Gerlach apparatus. Each atom emerges from the apparatus along one of three paths, corresponding to the component of angular momentum having the value $\hbar$, 0, $-\hbar$ along the direction of the magnetic field of the apparatus, which we take as the z axis. An atom which emerges along, say, the first path is *now* in the state u_{211}. This is a non-spherically symmetric function, and is related to a physically defined z axis.

When a large number of atoms have passed through the apparatus, they are found to be equally distributed among the three possibilities $\hbar$, 0, $-\hbar$. Therefore, the mean value of the component of angular momentum along the direction of the magnetic field is zero. The same result is obtained whatever the direction of the magnetic field, and we conclude that, for the atoms in the original system, the expectation value of the component of angular momentum along any axis is zero. This of course is to be expected, because the original system contained no physically defined axis.

We now consider how the state ψ of the atoms in the original system may be described. Suppose we set up a hypothetical set of Cartesian axes. Then we have three states u_{211}, u_{210}, u_{21-1} defined with respect to these axes. We know that if we do a Stern–Gerlach measurement with the magnetic field along the z axis we shall obtain an atom in one of these three states with equal probability. So ψ must be a mixture of equal parts of the three us. We write

$$\psi = (e_1 u_{211} + e_0 u_{210} + e_{-1} u_{21-1})/\sqrt{3}, \tag{2}$$

where

$$|e_1|^2 = |e_0|^2 = |e_{-1}|^2 = 1, \tag{3}$$

since ψ and the us are normalised.

The moduli of the three es are therefore unity. However, the phase relationships between them are arbitrary. This means that the functions are not properly known, which is a reflection of our lack of knowledge of the state of the system. To obtain complete knowledge (in the quantum sense) we must, in the present case, measure a component of angular momentum. The function ψ, representing the system *after* the measurement, will be free from this arbitrariness of the relative phases.

When we normally talk about the state ψ of a system, we mean one in which the relative phases are known. However, if it is desired to emphasise this aspect, the term *pure state* is sometimes used. The type of ψ we are discussing in the present problem is sometimes known as an *incoherent mixture* of states, or, more formally, as a *statistical ensemble* of states.

If we write such a state in the form

$$\psi = \frac{1}{\sqrt{n}} \sum_{i=1}^{i=n} e_i u_i, \tag{4}$$

then the arbitrariness in the phases means that whenever we have the product of two es we must take the statistical average, the value of which is given by

$$\langle e_i e_j^* \rangle = \delta_{ij}, \tag{5}$$

where δ_{ij} is the Kronecker delta. Thus

$$|\psi|^2 = \frac{1}{n} \sum_i |u_i|^2. \tag{6}$$

In other words, there is no interference between the different us, which accounts for the term *incoherent mixture*.

We return to the question of the angular dependence of $|\psi|^2$ for the original state $n = 2$, $l = 1$. From (1), (2), and (5)

$$|\psi|^2 = \tfrac{1}{3}|R_{21}(r)|^2\{|Y_{11}|^2 + |Y_{10}|^2 + |Y_{1-1}|^2\}. \tag{7}$$

Inserting the expressions for the Ys in Table 4.1 (p. 52) we have

$$\begin{aligned}\{|Y_{11}|^2 + |Y_{10}|^2 + |Y_{1-1}|^2\} &= \frac{3}{8\pi}\sin^2\theta + \frac{3}{4\pi}\cos^2\theta + \frac{3}{8\pi}\sin^2\theta \\ &= \frac{3}{4\pi}.\end{aligned} \tag{8}$$

So $|\psi|^2$ has the required spherical symmetry.

Eq. (8) is a special case of a general property of spherical harmonics, namely,

$$\sum_{m=-l}^{m=l} |Y_{lm}(\theta, \phi)|^2 = \frac{2l+1}{4\pi}. \tag{9}$$

The detailed form of the right-hand side is of no interest here – only the fact that it does not depend on θ or ϕ. So for a system of hydrogen atoms whose n and l quantum numbers have been determined – by measurement of the energy and resultant angular momentum – but for which no measurements have been made to determine the m quantum number, the square of the modulus of the state function is spherically symmetric.

4.9 We use the definitions and properties of the operators a and $a^\dagger$ on p. 54, and also the result that the energy of the eigenstate u_n is $E = (n + \frac{1}{2})\hbar\omega$.

(a) From (4.17)

$$x = (\hbar/2m\omega)^{1/2}(a + a^\dagger), \quad p = -\mathrm{i}(m\hbar\omega/2)^{1/2}(a - a^\dagger). \tag{1}$$

Thus

$$\langle x \rangle = (\hbar/2m\omega)^{1/2}(\langle a \rangle + \langle a^\dagger \rangle). \tag{2}$$

Now

$$\langle a \rangle = \int u_n^* a u_n \,\mathrm{d}x = \sqrt{n} \int u_n^* u_{n-1} \,\mathrm{d}x = 0, \tag{3}$$

since u_n and u_{n-1} are orthogonal. Similarly

$$\langle a^\dagger \rangle = 0. \tag{4}$$

Therefore

$$\langle x \rangle = \langle p \rangle = 0. \tag{5}$$

These two results are the same for the classical harmonic oscillator. If we average the displacement, taking account of sign, over one oscillation, the result is zero. Similarly the average of the momentum (or velocity) over one oscillation is zero.

(b) The expectation value of the potential energy is

$$\langle V \rangle = \tfrac{1}{2} m\omega^2 \langle x^2 \rangle. \tag{6}$$

From (1)

$$\langle x^2 \rangle = \frac{\hbar}{2m\omega} \langle (a + a^\dagger)(a + a^\dagger) \rangle. \tag{7}$$

Now

$$\langle aa \rangle = \int u_n^* aa u_n \,\mathrm{d}x = 0. \tag{8}$$

This follows because

$$aau_n = a(au_n) = a(\sqrt{n} u_{n-1}) = \sqrt{\{n(n-1)\}} u_{n-2}, \tag{9}$$

and u_n and u_{n-2} are orthogonal. Similarly $\langle a^\dagger a^\dagger \rangle = 0$. The term $\langle aa^\dagger \rangle$ is not zero. We have

$$aa^\dagger u_n = a(a^\dagger u_n) = a\{\surd(n+1)u_{n+1}\} = (n+1)u_n. \tag{10}$$

Therefore, since the u_n are normalised,

$$\langle aa^\dagger \rangle = n + 1. \tag{11}$$

Similarly

$$\langle a^\dagger a \rangle = n. \tag{12}$$

Inserting (11) and (12) in (7) gives

$$\langle x^2 \rangle = \frac{\hbar}{m\omega}(n + \tfrac{1}{2}). \tag{13}$$

Therefore from (6)

$$\langle V \rangle = \tfrac{1}{2}(n + \tfrac{1}{2})\hbar\omega = \tfrac{1}{2}E, \tag{14}$$

where E is the total energy. The expectation value of the kinetic energy T is obtained immediately from the relation

$$\langle T \rangle = E - \langle V \rangle = \tfrac{1}{2}E. \tag{15}$$

The result $\langle T \rangle = \langle V \rangle$ may be obtained from the virial theorem. For the harmonic oscillator, $V \propto x^2$, so $n = 2$ in (4.6.19), giving the stated result.

(c) Δ_x, the uncertainty in x, is defined by

$$\Delta_x^2 = \langle(\langle x \rangle - x)^2\rangle = \langle x^2 \rangle, \quad \text{since} \quad \langle x \rangle = 0. \tag{16}$$

From (6) and (14)

$$\Delta_x^2 = \langle x^2 \rangle = \frac{2}{m\omega^2}\langle V \rangle = \frac{E}{m\omega^2}. \tag{17}$$

Similarly

$$\Delta_p^2 = \langle p^2 \rangle = 2m\langle T \rangle = mE. \tag{18}$$

From (17) and (18)

$$\Delta_x\Delta_p = \frac{E}{\omega} = (n + \tfrac{1}{2})\hbar. \tag{19}$$

(d) The annihilation operator a operating on the ground-state eigenfunction gives zero, i.e. $au_0 = 0$. Now

$$a = (2m\hbar\omega)^{-1/2}(m\omega x + ip), \quad \text{and} \quad ip = \hbar\frac{\mathrm{d}}{\mathrm{d}x}. \tag{20}$$

Therefore

$$\left(m\omega x + \hbar\frac{\mathrm{d}}{\mathrm{d}x}\right)u_0 = 0, \tag{21}$$

i.e.

$$\frac{\mathrm{d}u_0}{\mathrm{d}x} = -\frac{m\omega}{\hbar}xu_0, \tag{22}$$

the solution of which is

$$u_0 = A\exp\left(-\frac{m\omega}{2\hbar}x^2\right), \tag{23}$$

where A is a normalising constant.

To establish the parity of the state u_n, we note first that the operator $a^\dagger$ has odd parity, and therefore, from the relation $a^\dagger u_n = \sqrt{(n+1)}u_{n+1}$, u_n and u_{n+1} have opposite parity. From (23) u_0 has even parity. It therefore follows that the parity of u_n is even for even n, and odd for odd n.

Comments

(1) For the ground state, (19) becomes

$$\Delta_x\Delta_p = \tfrac{1}{2}\hbar. \tag{24}$$

We prove in Problem 10.1 that this value of $\Delta_x\Delta_p$, the minimum allowed by the uncertainty principle, is obtained only with a Gaussian state function, which is consistent with the result in (23).

(2) The potential $V(x)$ of a harmonic oscillator is proportional to x^2 and therefore has even parity. The energy values are non-degenerate. The results of the problem provide another illustration that under these conditions the solutions of the Schrödinger equation have a definite parity.

4.10 (a) Put

$$H = H_x + H_y + H_z, \tag{1}$$

where

$$H_x = -\frac{\hbar^2}{2m}\frac{\partial^2}{\partial x^2} + \tfrac{1}{2}m\omega^2x^2, \tag{2}$$

and similarly for H_y, H_z. The operators H_x, H_y, H_z differ only in the labelling of the coordinates. Put

$$U(x, y, z) = u_1(x)u_2(y)u_3(z), \tag{3}$$

where $u_1(x)$ is a solution of

$$H_x u_1(x) = E_1 u_1(x), \tag{4}$$

and $u_2(y)$, $u_3(z)$ are solutions of corresponding equations. Then $U(x, y, z)$ satisfies

$$HU(x, y, z) = EU(x, y, z), \tag{5}$$

where

$$E = E_1 + E_2 + E_3. \tag{6}$$

This follows because when the H_x term in H operates on

$$u_1(x)u_2(y)u_3(z),$$

it treats $u_2(y)u_3(z)$ as a constant. Therefore from (4)

$$H_x u_1(x)u_2(y)u_3(z) = E_1 u_1(x)u_2(y)u_3(z), \tag{7}$$

and similarly for the H_y and H_z terms.

But, from (2) and (4), $u_1(x)$ is an energy eigenfunction for a one-dimensional harmonic oscillator. The energy has the value

$$E_1 = (n_1 + \tfrac{1}{2})\hbar\omega, \tag{8}$$

where n_1 is zero or a positive integer. Similarly for E_2 and E_3. The energy of the eigenstate $u_{n1}(x)u_{n2}(y)u_{n3}(z)$ is therefore

$$E = E_1 + E_2 + E_3 = (n + \tfrac{3}{2})\hbar\omega, \tag{9}$$

where

$$n = n_1 + n_2 + n_3 \tag{10}$$

is zero or a positive integer.

(b) The eigenfunction

$$U(x, y, z) = u_{n1}(x)u_{n2}(y)u_{n3}(z) \tag{11}$$

is specified by the trio of integers (n_1, n_2, n_3). The ground state has $n = 0$, which must correspond to $(0, 0, 0)$. The next energy level is $n = 1$, which can be obtained from each of the states

$$(1, 0, 0), (0, 1, 0), (0, 0, 1).$$

The $n = 2$ energy level can be obtained from

$$(2, 0, 0), (0, 2, 0), (0, 0, 2), (0, 1, 1), (1, 0, 1), (1, 1, 0).$$

Thus the three levels $n = 0, 1, 2$ have degeneracies 1, 3, 6.

To obtain the degeneracy of the state n, we need to find the number of different trios of integers whose sum is n. We may do this as follows. Suppose we have n balls arranged in a line. We take a partition and put it at either end of the line of balls, or between any two of the balls. There are $n + 1$ possible positions. We now take a second partition and place it similarly in the line, this time treating the first partition as a ball. The second partition can thus go into $n + 2$ possible positions. The number of position combinations for the two partitions is thus $\frac{1}{2}(n + 1)(n + 2)$. The factor $\frac{1}{2}$ comes from the fact that the partitions are to be regarded as indistinguishable, and may be exchanged. Each position combination corresponds to a set of integers whose sum is n. The number of balls to the left of the first partition gives n_1, the number between the two partitions gives n_2, and the number to the right of the second partition gives n_3.

(c) If n is even, then, from (10), either n_1, n_2, n_3 are all even, or one of them is even and the other two are odd. From the results of Problems 3.2(b) and 4.9(d), it follows that the parity of the states of the level n is even. Similarly, if n is odd, the parity of the states is odd.

5

Matrices, spin, addition of angular momentum

Summary of theory

1 Matrices

Let $\hat{A}$ be an operator, and u_i ($i = 1$ to N) a set of orthonormal functions. The matrix $\tilde{A}$ corresponding to the operator $\hat{A}$ on the basis of the functions u_i is defined by

$$A_{mn} = \int_{\substack{\text{all}\\ \text{space}}} u_m^* \hat{A} u_n \, \mathrm{d}\tau. \tag{1}$$

$\tilde{A}$ is a square matrix of dimensions $N \times N$. The following important results follow from the definition.

(i) $$\hat{A} u_j = \sum_k A_{kj} u_k. \tag{2}$$

(ii) Any arithmetical relation that holds between operators holds between the corresponding matrices, e.g.

$$\text{if} \quad \hat{C} = \hat{A} + \hat{B}, \quad \text{then} \quad \tilde{C} = \tilde{A} + \tilde{B}, \tag{3}$$

$$\text{if} \quad \hat{C} = \hat{A}\hat{B}, \quad \text{then} \quad \tilde{C} = \tilde{A}\tilde{B}. \tag{4}$$

The same set of basis functions must be used throughout.

(iii) The eigenvalues of the matrix $\tilde{A}$ are the same as the eigenvalues of the operator $\hat{A}$. The eigenvector of $\tilde{A}$ corresponding to an eigenvalue λ is the corresponding eigenfunction of $\hat{A}$. Thus if $\tilde{A}X = \lambda X$, where X is the column vector

$$\begin{bmatrix} X_1 \\ X_2 \\ \cdots \\ X_N \end{bmatrix},$$

then $\hat{A}\phi = \lambda\phi$, where $\phi = \sum_i X_i u_i$.

(iv) If the basis functions u_i are eigenfunctions of $\hat{A}$ with eigenvalues λ_i, then $\widetilde{A}$ is a diagonal matrix with diagonal elements equal to λ_i, i.e. it has the form

$$\begin{bmatrix} \lambda_1 & 0 & 0 & \cdots \\ 0 & \lambda_2 & 0 & \cdots \\ 0 & 0 & \lambda_3 & \cdots \\ \cdots & & & \end{bmatrix}.$$

2 Dirac notation

A state function ϕ is written as $|\phi\rangle$, and ϕ^* is written as $\langle\phi|$. The Dirac notation is very convenient if we are dealing with a known set of functions u_i, and wish to emphasise relations between functions with different values of i; the symbol u is then omitted. The notation lends itself to a compact way of expressing the integrals over all space that constantly arise in the theory. Some examples follow.

standard	Dirac
u_j	$\lvert j\rangle$
u_i^*	$\langle i\rvert$
$\int_{\text{all space}} u_i^* A u_j \, d\tau$	$\langle i\lvert A\rvert j\rangle$

3 Spin angular momentum

The operators corresponding to the x, y, z components of spin angular momentum are denoted by S_x, S_y, S_z. The operator $S^2 = S_x^2 + S_y^2 + S_z^2$ corresponds to the square of the magnitude of the spin angular momentum. The commutation relations for S^2, S_x, S_y, S_z are the same as for the orbital angular momentum operators. The pattern of the eigenvalues of the common eigenfunctions of the operators S^2, S_z is the same as for orbital angular momentum, but the quantum number l, denoted by s for spin angular momentum, can take half-integral values also, i.e.

$$s = 0, 1/2, 1, 3/2, 2, 5/2, \ldots \tag{5}$$

For given s the quantum number m, denoted by m_s, takes values

$$m_s = s, s-1, s-2, \ldots -(s-1), -s. \tag{6}$$

For a single electron, s takes only the value $\frac{1}{2}$, with $m_s = \pm\frac{1}{2}$. The two eigenfunctions are denoted by α and β. Thus

$$S^2\alpha = s(s+1)\hbar^2\alpha, \quad S^2\beta = s(s+1)\hbar^2\beta, \quad s = \tfrac{1}{2}, \tag{7}$$

$$S_z\alpha = \tfrac{1}{2}\hbar\alpha, \qquad S_z\beta = -\tfrac{1}{2}\hbar\beta. \tag{8}$$

4 Abstract eigenfunctions and Hilbert space

Unlike the operators and eigenfunctions for orbital angular momentum, which are specific functions of the space variables, the operators and eigenfunctions for spin do not have a functional form in terms of space variables. The spin operators are defined only by their commutation properties, and the eigenfunctions only by relations such as (7) and (8). The eigenfunctions are said to be unit vectors in an abstract space, known as *Hilbert space*. (In fact all eigenfunctions, including those that are functions of space variables, may be considered as unit vectors in Hilbert space.) Do not be put off by this highbrow language – the procedures, which we here outline, are quite straightforward.

Consider the case of spin $s = \frac{1}{2}$, where there are two eigenfunctions or unit vectors, α and β, so the Hilbert space for this case is two-dimensional. The spin state $\gamma = a\alpha + b\beta$, where a and b are numbers, may be represented as a vector in the space with coordinates a, b. The numbers a and b are in general complex.

Since α and β are not functions of space, we cannot use (1) to calculate a matrix representation for the spin operators. Instead we define a quantity, known as the *scalar product* of two state vectors γ_1 and γ_2, and use the Dirac notation to denote it by $\langle\gamma_1|\gamma_2\rangle$. The definition is an operational one, that is to say, it gives the rules for calculating the scalar product, which is, in general, a complex number. The scalar product of α (or β) with itself is unity, while the scalar product of α and β is zero. Thus

$$\langle\alpha|\alpha\rangle = \langle\beta|\beta\rangle = 1, \quad \langle\alpha|\beta\rangle = \langle\beta|\alpha\rangle = 0. \tag{9}$$

In the usual language we say that α and β are normalised and orthogonal to each other. The scalar product of $\gamma_1 = a_1\alpha + b_1\beta$ and $\gamma_2 = a_2\alpha + b_2\beta$ is then defined to be

$$\begin{aligned}\langle\gamma_1|\gamma_2\rangle &= \langle a_1\alpha + b_1\beta|a_2\alpha + b_2\beta\rangle \\ &= a_1^*a_2\langle\alpha|\alpha\rangle + a_1^*b_2\langle\alpha|\beta\rangle + b_1^*a_2\langle\beta|\alpha\rangle + b_1^*b_2\langle\beta|\beta\rangle \\ &= a_1^*a_2 + b_1^*b_2. \end{aligned}\tag{10}$$

You can see that the definition follows from (9), with the extra condition that the numbers in the first vector are replaced by their complex conjugates. The matrix element A_{mn} for a spin operator $\hat{A}$ is defined by

$$A_{mn} = \langle m|\hat{A}n\rangle, \tag{11}$$

where m and n are α or β. So the matrix has dimensions 2×2. Postulate 4 – p. 9 – is similarly modified for spin operators and states. If $\hat{S}$ is the operator corresponding to a spin observable σ, then the expectation value of σ for the state γ is

$$\langle \hat{S}\rangle = \langle \gamma|\hat{S}\gamma\rangle. \tag{12}$$

The method is readily extended to higher spin values. Thus for $s = 1$, there are $2s + 1 = 3$ normalised and mutually orthogonal spin eigenfunctions or state vectors. The Hilbert space is three-dimensional, and the matrix representing a spin operator has dimensions 3×3, and so on. See also the comment at the end of the solution to Problem 5.7.

5 Addition of angular momentum

(a) Definition of operators

Classically the resultant of two angular momenta $\mathbf{J}_1$ and $\mathbf{J}_2$ is the vector sum

$$\mathbf{J} = \mathbf{J}_1 + \mathbf{J}_2. \tag{13}$$

Each component of $\mathbf{J}$ is the sum of the corresponding components of $\mathbf{J}_1$ and $\mathbf{J}_2$. Thus

$$J_x = J_{1x} + J_{2x}, \quad J_y = J_{1y} + J_{2y}, \quad J_z = J_{1z} + J_{2z}. \tag{14}$$

The square of the magnitude of $\mathbf{J}$ is given by

$$J^2 = J_1^2 + J_2^2 + 2\mathbf{J}_1\cdot\mathbf{J}_2 \tag{15}$$

$$= J_1^2 + J_2^2 + 2(J_{1x}J_{2x} + J_{1y}J_{2y} + J_{1z}J_{2z}). \tag{16}$$

J_{1x}, J_{2x}, etc. and J_1^2 and J_2^2 are represented by operators as in the previous treatment of angular momentum. The commutation relations for the J_1 operators are, as before,

$$[J_1^2, J_{1x}] = [J_1^2, J_{1y}] = [J_1^2, J_{1z}] = 0, \tag{17}$$

$$[J_{1x}, J_{1y}] = i\hbar J_{1z}, \quad \text{and cyclic permutations.} \tag{18}$$

Similarly for the J_2 operators. Any 1 operator, i.e. J_1^2, J_{1x}, etc. commutes with any 2 operator, J_2^2, J_{2x}, etc. The angular momenta $\mathbf{J}_1$ and $\mathbf{J}_2$ may be orbital or spin.

To find the operators for the components of $\mathbf{J}$ and for J^2, we use (14) and (16), interpreting the quantities on the right-hand sides as operators.

It is easily verified that the commutation relations between the operators J^2, J_x, J_y, J_z are the same as (17) and (18). Thus J^2 commutes with each of its components J_x, J_y, J_z, and the commutation relations between the components are

$$[J_x, J_y] = i\hbar J_z, \quad \text{and so on.} \tag{19}$$

(b) Eigenfunctions and eigenvalues

Let $\phi^{(1)}_{j_1 m_1}$ be an eigenfunction of the operators J_1^2 and J_{1z}. Then from the previous results on orbital and spin angular momentum

$$J_1^2 \phi^{(1)}_{j_1 m_1} = j_1(j_1 + 1)\hbar^2 \phi^{(1)}_{j_1 m_1}, \tag{20}$$

$$J_{1z} \phi^{(1)}_{j_1 m_1} = m_1 \hbar \phi^{(1)}_{j_1 m_1}. \tag{21}$$

j_1 takes values $0, \frac{1}{2}, 1, \frac{3}{2}, 2, \ldots$ (The half-integral values are excluded for orbital angular momentum.) For given j_1, m_1 takes values j_1, $j_1 - 1$, $j_1 - 2, \ldots -j_1$.

Since J^2 and J_z commute, they have a common set of eigenfunctions. Denote them by Φ_{jm}. The pattern of eigenvalues for these functions is the same as for the component angular momenta. The eigen equations are

$$J^2 \Phi_{jm} = j(j + 1)\hbar^2 \Phi_{jm}, \tag{22}$$

$$J_z \Phi_{jm} = m\hbar \Phi_{jm}, \tag{23}$$

where j takes values $0, \frac{1}{2}, 1, \frac{3}{2}, 2, \ldots$, and, for each j, the values of m range from j to $-j$ in integral steps. For a particular pair of j_1, j_2 values, the values of j are further restricted to

$$j = j_1 + j_2, \quad j_1 + j_2 - 1, \quad j_1 + j_2 - 2, \quad \ldots |j_1 - j_2|. \tag{24}$$

(c) Two sets of mutually commuting operators

The commutation relations between the operators are summarised by the following scheme.

$$\begin{array}{ccccc} & J^2 & & \mathbf{J}_1\cdot\mathbf{J}_2 & \\ J_1^2 & & J_2^2 & & J_z \\ & J_{1z} & & J_{2z} & \end{array} \tag{25}$$

An operator commutes with any other operator on the same or adjacent line, but an operator on the top line does not commute with an operator on the bottom line.

The eigenfunctions of the mutually commuting set of operators J_1^2, J_2^2,

J_{1z}, J_{2z} are the product functions $\phi^{(1)}_{j_1m_1}\phi^{(2)}_{j_2m_2}$. This follows because J_1^2 and J_{1z} operate only on variables (1), and treat $\phi^{(2)}_{j_2m_2}$ as a constant. Therefore from (20)

$$J_1^2\phi^{(1)}_{j_1m_1}\phi^{(2)}_{j_2m_2} = j_1(j_1+1)\hbar^2\phi^{(1)}_{j_1m_1}\phi^{(2)}_{j_2m_2}. \tag{26}$$

Similarly for J_{1z}, J_2^2, and J_{2z}.

Denote the eigenfunctions of the mutually commuting set of operators J_1^2, J_2^2, J^2, J_z by $\Phi_{j_1j_2jm}$. Thus

$$J_1^2\Phi_{j_1j_2jm} = j_1(j_1+1)\hbar^2\Phi_{j_1j_2jm}, \quad J_2^2\Phi_{j_1j_2jm} = j_2(j_2+1)\hbar^2\Phi_{j_1j_2jm},$$
$$J^2\Phi_{j_1j_2jm} = j(j+1)\hbar^2\Phi_{j_1j_2jm}, \quad J_z\Phi_{j_1j_2jm} = m\hbar\Phi_{j_1j_2jm}. \tag{27}$$

The functions $\Phi_{j_1j_2jm}$ depend on both the (1) and (2) variables.

For a given value of j_1, there are $2j_1+1$ values of m_1, and similarly for j_2. Thus there are $(2j_1+1)(2j_2+1)$ product eigenfunctions $\phi^{(1)}_{j_1m_1}\phi^{(2)}_{j_2m_2}$ corresponding to a given j_1, j_2 combination. It is readily shown that, for the j values given in (24), each with its set of $2j+1$ values of m, there are also $(2j_1+1)(2j_2+1)$ of the $\Phi_{j_1j_2jm}$ eigenfunctions. The two sets of eigenfunctions are not independent. We can express any eigenfunction $\Phi_{j_1j_2jm}$ as a linear combination of the product eigenfunctions $\phi^{(1)}_{j_1m_1}\phi^{(2)}_{j_2m_2}$, and vice-versa. The coefficients in these expressions, known as *Clebsch–Gordan* coefficients, depend on the values of j_1, j_2, m_1, m_2, j, m, and have been calculated once and for all for the various low-value combinations of the permitted integer and half-integers. A simplifying feature in the calculation is that $\Phi_{j_1j_2jm}$ is an eigenfunction of J_z with eigenvalue m, and $\phi^{(1)}_{j_1m_1}\phi^{(2)}_{j_2m_2}$ is also an eigenfunction of J_z with eigenvalue m_1+m_2. Therefore, in representing $\Phi_{j_1j_2jm}$ as a linear combination of the $\phi^{(1)}_{j_1m_1}\phi^{(2)}_{j_2m_2}$ functions, we do not need all the $(2j_1+1)(2j_2+1)$ terms, but only those for which $m_1+m_2=m$. You can do the calculation yourself for the simple cases $j_1=j_2=\frac{1}{2}$, and $j_1=1$, $j_2=\frac{1}{2}$ (Problem 5.7).

(d) Eigenfunctions of the Hamiltonian

Any function $\psi_{j_1j_2}$, which is an eigenfunction of the operators J_1^2 and J_2^2, with eigenvalues j_1, j_2, can be represented in terms of either the $\phi^{(1)}_{j_1m_1}\phi^{(2)}_{j_2m_2}$ functions, or the $\Phi_{j_1j_2jm}$ functions. Since we are usually interested in states that are eigenfunctions of the Hamiltonian, our choice of representation is guided by the form of the Hamiltonian, in particular by the presence of angular momentum terms.

Angular momentum operators arise in the Hamiltonian from magnetic forces. A particle with angular momentum has a magnetic dipole moment proportional to its angular momentum, so the operator corresponding to the magnetic dipole moment is some constant times the angular momentum operator. The application of an external magnetic field **B** along the z axis to a system with two sources of angular momentum $\mathbf{J}_1$ and $\mathbf{J}_2$, gives rise to terms in the Hamiltonian proportional to J_{1z} and J_{2z}. In addition there will be interactions between the magnetic dipoles themselves, which will give a contribution to the Hamiltonian proportional to $\mathbf{J}_1 \cdot \mathbf{J}_2$.

The Hamiltonian may be written in the form

$$H = H^0 + \eta \mathbf{J}_1 \cdot \mathbf{J}_2 + \beta_1 J_{1z} + \beta_2 J_{2z}. \tag{28}$$

H^0 is the part of the Hamiltonian that represents the kinetic energy and the potential energy due to a central force. It contains no magnetic forces. The term $\eta \mathbf{J}_1 \cdot \mathbf{J}_2$ represents internal magnetic forces such as spin–orbit coupling; η is a constant that depends on various atomic parameters. The terms $\beta_1 J_{1z}$ and $\beta_2 J_{2z}$ are due to an external magnetic field **B**; the constants β_1 and β_2 are proportional to B.

The term H^0 commutes with all the angular momentum operators, so it plays no part in the discussion. We may distinguish three cases.

(1) No external magnetic field

The Hamiltonian $H = H^0 + \eta \mathbf{J}_1 \cdot \mathbf{J}_2$ commutes with the operators J_1^2, J_2^2, J^2, J_z, but not with J_{1z} or J_{2z}. So an eigenfunction of the Hamiltonian has the form $\Phi_{j_1 j_2 jm}$. That is to say, if we choose to represent the eigenfunction in terms of the $\Phi_{j_1 j_2 jm}$ functions, we shall have only a single term in the representation. If we choose the $\phi^{(1)}_{j_1 m_1} \phi^{(2)}_{j_2 m_2}$ representation, we shall, in general, have several terms in the expansion. Naturally we would choose the former.

(2) Very large external magnetic field

By this we mean that the magnetic field is so large that the terms $\beta_1 J_{1z}$ and $\beta_2 J_{2z}$ completely dominate the term $\eta \mathbf{J}_1 \cdot \mathbf{J}_2$. If we neglect the latter term, the Hamiltonian is $H = H^0 + \beta_1 J_{1z} + \beta_2 J_{2z}$, which commutes with J_1^2, J_2^2, J_{1z}, J_{2z}, but not with J^2. So an eigenfunction of the Hamiltonian has the form $\phi^{(1)}_{j_1 m_1} \phi^{(2)}_{j_2 m_2}$. We choose this representation because it contains only a single term. The $\Phi_{j_1 j_2 jm}$ representation would, in general, contain several terms.

(3) Intermediate magnetic field

For intermediate values of the magnetic field, the Hamiltonian has the general form of (28). It still commutes with J_1^2, J_2^2, and J_z, but not with J_{1z}, J_{2z}, or J^2. Therefore, whichever of the two representations is used for an eigenfunction of the Hamiltonian, it will, in general, contain more than one term.

6 Summary – angular momentum in general

Let J_x, J_y, J_z be the operators corresponding to the components of some angular momentum. It may be an orbital, or a spin angular momentum, or a resultant obtained by adding two angular momenta, which may themselves be either orbital or spin or resultant angular momenta. Let $J^2 = J_x^2 + J_y^2 + J_z^2$. Then in all cases the same commutation relations

$$[J^2, J_x] = [J^2, J_y] = [J^2, J_z] = 0, \tag{29}$$

$$[J_x, J_y] = \mathrm{i}\hbar J_z, \text{ etc.} \tag{30}$$

are obeyed. And in all cases the same pattern of eigenvalues for the common eigenfunctions of J^2 and J_z holds, namely

$$J^2\phi_{jm} = j(j+1)\hbar^2\phi_{jm}, \quad J_z\phi_{jm} = m\hbar\phi_{jm}, \tag{31}$$

where j takes integral values for orbital angular momentum, and integral or half-integral values for spin or combination of spin and orbital angular momentum. For a given j, m takes values from j to $-j$ in integral steps. In all cases we can define raising and lowering operators by

$$J_+ = J_x + \mathrm{i}J_y, \quad J_- = J_x - \mathrm{i}J_y, \tag{32}$$

which satisfy the relations

$$J_+\phi_{jm} = \{j(j+1) - m(m+1)\}^{1/2}\hbar\phi_{j,m+1}, \tag{33}$$

$$J_-\phi_{jm} = \{j(j+1) - m(m-1)\}^{1/2}\hbar\phi_{j,m-1}. \tag{34}$$

Problems

5.1 $\tilde{A}$, $\tilde{B}$, $\tilde{C}$ are matrices generated on the same set of basis functions by the operators A, B, C. Show that

(a) if $C = A + B$, then $\tilde{C} = \tilde{A} + \tilde{B}$,
(b) if $C = AB$, then $\tilde{C} = \tilde{A}\tilde{B}$.

5.2 (a) Using the functions ϕ_1, ϕ_0, ϕ_{-1} in Problem 4.2 as basis, calculate the matrices $\tilde{L}_x$, $\tilde{L}_y$, $\tilde{L}_z$, $\tilde{L}^2$.

(b) Show that the eigenvalues of $\tilde{L}_x$ and $\tilde{L}^2$ have the expected values.

5.3 The Pauli spin operators σ_x, σ_y, σ_z are defined in terms of the spin angular momentum operators S_x, S_y, S_z by $\sigma_x = 2S_x/\hbar$, and similarly for y and z. Consider the case $s = \frac{1}{2}$, and denote the normalised eigenfunctions of S_z by α and β. Use the relations for the raising and lowering operators (5.33) and (5.34) to prove the following results.

(a)
$$\sigma_x\alpha = \beta, \quad \sigma_y\alpha = \mathrm{i}\beta, \quad \sigma_z\alpha = \alpha,$$
$$\sigma_x\beta = \alpha, \quad \sigma_y\beta = -\mathrm{i}\alpha, \quad \sigma_z\beta = -\beta.$$

(b) The normalised eigenfunctions of σ_x, σ_y, σ_z are

operator	eigenfunctions	
σ_x	$(\alpha + \beta)/\sqrt{2}$	$(\alpha - \beta)/\sqrt{2}$
σ_y	$(\alpha + \mathrm{i}\beta)/\sqrt{2}$	$(\alpha - \mathrm{i}\beta)/\sqrt{2}$
σ_z	α	β

The eigenvalues are $+1$ for the eigenfunctions in the left-hand column, and -1 for those in the right-hand column.

(c) The matrices of σ_x, σ_y, σ_z on the basis of α and β are

$$\tilde{\sigma}_x = \begin{bmatrix} 0 & 1 \\ 1 & 0 \end{bmatrix}, \quad \tilde{\sigma}_y = \begin{bmatrix} 0 & -\mathrm{i} \\ \mathrm{i} & 0 \end{bmatrix}, \quad \tilde{\sigma}_z = \begin{bmatrix} 1 & 0 \\ 0 & -1 \end{bmatrix}.$$

5.4 P is a beam of atoms with spin quantum number $\frac{1}{2}$ and zero orbital angular momentum, all with angular momentum $+\hbar/2$ along the x axis. Q is a beam of similar but unpolarised atoms.

(a) What is the spin state function of P in terms of α and β, the eigenfunctions of S_z?

(b) If the two beams are passed separately through a Stern–Gerlach apparatus with its magnetic field along the z axis, is there any difference between the emerging beams in the two cases?

(c) How could the difference between P and Q be detected experimentally?

5.5 The beam Q in the last problem is an incoherent mixture of the states α and β in equal proportions. Its spin state function may therefore be written as

$$\psi = (e_1\alpha + e_2\beta)\sqrt{2},$$

where e_1 and e_2 are to be regarded as complex numbers of modulus unity with random relative phases, i.e. they satisfy the relations

$$|e_1|^2 = |e_2|^2 = 1, \quad \langle e_1^* e_2 \rangle = \langle e_2^* e_1 \rangle = 0,$$

where the brackets $\langle\ \rangle$ indicate the average over all values of the relative phase. By expressing ψ in terms of the eigenfunctions of S_x, show that this state function gives the required physical result, namely, that if an unpolarised beam of the atoms is passed through a Stern–Gerlach apparatus with its magnetic field in the x direction, the two emerging beams contain equal numbers of atoms.

5.6 A beam of atoms with spin quantum number $\frac{1}{2}$ and zero orbital angular momentum passes through a Stern–Gerlach magnet whose magnetic field is along a direction D at an angle θ to the z axis. The emerging beam with spins along D is passed through a second Stern–Gerlach magnet with its magnetic field along the z axis. Show that in the two beams that emerge from the second magnet the numbers of atoms with spins parallel and anti-parallel to the z axis are in the ratio $\cos^2\frac{1}{2}\theta : \sin^2\frac{1}{2}\theta$.

5.7 (a) Let **S** be the operator for the resultant spin angular momentum of two electrons, and S_z its z component. If Φ_{SM} is an eigenfunction of S^2 and S_z with respective eigenvalues $S(S+1)\hbar^2$ and $M\hbar$, derive the expression for each Φ_{SM} in terms of the product functions $\alpha\alpha$, $\alpha\beta$, $\beta\alpha$, $\beta\beta$, where the first α or β refers to electron 1, and the second to electron 2.

(b) Consider the addition of an orbital angular momentum **L** and a spin angular momentum **S** for the case $l = 1$, $s = \frac{1}{2}$. The eigenfunctions of the operators L^2, S^2, L_z, S_z are products of the space functions ϕ_1, ϕ_0, ϕ_{-1} and the spin functions α, β. Derive the eigenfunctions Φ_{jm_j} of the operators L^2, S^2, J^2, J_z in terms of the first set of eigenfunctions.
[Hint: In both cases start with the Φ function in which S, M (or j, m_j) have their maximum values, and apply the appropriate lowering operators to both sides of the equation.]

5.8 A beam of atoms in the state $l = 1$, $s = \frac{1}{2}$, $j = \frac{3}{2}$ passes through a Stern-Gerlach apparatus in which the magnetic field is small compared to E_0/μ_B, where E_0 is the spin–orbit energy, (i.e. the energy separation of the states $l = 1$, $s = \frac{1}{2}$, $j = \frac{3}{2}$ and $l = 1$, $s = \frac{1}{2}$, $j = \frac{1}{2}$), and μ_B is the Bohr magneton. The four emerging beams, which contain equal numbers of atoms, are separated and, continuing in the same direction, each passes through a separate Stern–Gerlach apparatus in which the magnetic fields are large compared to E_0/μ_B. The direction of all five magnetic fields is the same and at right angles to the beam.

Into how many beams is each of the four beams further split, and what are the relative numbers of atoms in the final beams?

5.9 This problem is a calculation of the hyperfine structure of the energy levels of hydrogen in a magnetic field. Consider a hydrogen atom in the 1*s* state. Denote the *z* component of the Pauli spin operator for the electron by σ_{ez}, and its eigenstates by α_e and β_e, with corresponding notation for the proton nucleus. In the presence of a uniform magnetic field **B** in the *z* direction, the magnetic terms in the Hamiltonian may be written as

$$H = B(\mu_e \sigma_{ez} + \mu_p \sigma_{pz}) + W \boldsymbol{\sigma}_e \cdot \boldsymbol{\sigma}_p,$$

where the vector components of $\boldsymbol{\sigma}$ are the Pauli spin operators, and W is a constant. The first term represents the interaction between the magnetic dipole moments $\boldsymbol{\mu}_e$ and $\boldsymbol{\mu}_p$ of the electron and proton with the magnetic field, and the second term the magnetic interaction between the dipoles. The numerical values of the magnetic dipole moments are $\mu_e = \mu_B$, and $\mu_p = 2.79\mu_N$, where μ_N is the nuclear magneton. Since $\mu_e \gg \mu_p$, the term in μ_p in the Hamiltonian may be neglected.

Show the following.

(a) For $B = 0$, the eigenstates of H are $\alpha\alpha$, $\beta\beta$, $(\alpha\beta + \beta\alpha)/\sqrt{2}$, with energy $+W$, and $(\alpha\beta - \beta\alpha)/\sqrt{2}$ with energy $-3W$. (The first α or β in each product refers to the electron and the second to the proton.)

(b) For general values of B, the energy values are

$$W \pm \varepsilon, \; -W \pm (4W^2 + \varepsilon^2)^{1/2},$$

where $\varepsilon = \mu_e B$.

(c) Sketch the energy values as a function of B, labelling the curves with as much information as possible about the angular momentum of the states.

(d) Estimate the value of the hyperfine separation $4W$ (i.e. the separation of the two energy values at $B = 0$) from a classical model of two magnetic dipoles, taking the simple case in which the dipole moments are parallel to the line joining them. Compare your result with the observed value $4W = 5.9\ \mu\text{eV}$.

5.10 This problem demonstrates the Zeeman effect for a one-electron atom with the valency electron in a state with $l = 1$. In the presence of a uniform magnetic field $\mathbf{B}$ along the z axis, the magnetic terms in the Hamiltonian may be written as

$$H = \frac{\mu_B B}{\hbar}(L_z + 2S_z) + \frac{2W}{\hbar^2}\mathbf{L}\cdot\mathbf{S}.$$

$\mathbf{L}$ and $\mathbf{S}$ are the operators for the orbital and spin angular momentum, and L_z and S_z are their z components. μ_B is the Bohr magneton, and W is a constant of order $\mu_0 Z^4 e^2 \hbar^2 / 4\pi m_e^2 a_0^3$, where Ze is the effective charge on the nucleus, m_e is the mass of the electron, and a_0 is the Bohr radius. The first term in H is the energy due to the interaction of the orbital and spin magnetic dipole moments of the electron with the external magnetic field $\mathbf{B}$, and the second term is the spin–orbit interaction, i.e. the interaction between the orbital and spin magnetic dipoles themselves. Denote the three eigenfunctions of L_z by ϕ_1, ϕ_0, ϕ_{-1}, the eigenfunctions of S_z by α and β, and the product eigenfunctions by

$$\chi_1 = \phi_1\alpha, \quad \chi_3 = \phi_0\alpha, \quad \chi_5 = \phi_{-1}\alpha,$$

$$\chi_2 = \phi_1\beta, \quad \chi_4 = \phi_0\beta, \quad \chi_6 = \phi_{-1}\beta.$$

Show the following.

(a) $\mathbf{L}\cdot\mathbf{S} = \frac{1}{2}(L_+S_- + L_-S_+) + L_zS_z$,

where L_+ and L_- are the raising and lowering operators for the orbital angular momentum, and S_+ and S_- the corresponding operators for the spin.

(b)
$$H\chi_1 = c_{11}\chi_1, \qquad H\chi_6 = c_{66}\chi_6,$$
$$H\chi_2 = c_{22}\chi_2 + c_{23}\chi_3, \qquad H\chi_3 = c_{32}\chi_2 + c_{33}\chi_3,$$
$$H\chi_4 = c_{44}\chi_4 + c_{45}\chi_5, \qquad H\chi_5 = c_{54}\chi_4 + c_{55}\chi_5,$$

where

$$c_{11} = 2\varepsilon + W, \qquad c_{66} = -2\varepsilon + W,$$
$$c_{22} = c_{55} = -W, \qquad c_{33} = -c_{44} = \varepsilon,$$
$$c_{23} = c_{32} = c_{45} = c_{54} = \sqrt{2}W, \qquad \varepsilon = \mu_B B.$$

(c) The eigenvalues E of the Hamiltonian have the values

$$W \pm 2\varepsilon, \quad \tfrac{1}{2}[(\varepsilon - W) \pm \{(\varepsilon + W)^2 + 8W^2\}^{1/2}],$$
$$\tfrac{1}{2}[-(\varepsilon + W) \pm \{(\varepsilon - W)^2 + 8W^2\}^{1/2}].$$

(d) The variation of E with B, (i) for $\varepsilon \ll W$, and (ii) for $\varepsilon \gg W$, is in accord with the vector model for the case $l = 1$, $s = \frac{1}{2}$.

Solutions

5.1 (a) From (5.1)

$$C_{ij} = \int u_i^*(A + B)u_j \, d\tau = \int u_i^* A u_j \, d\tau + \int u_i^* B u_j \, d\tau$$

$$= A_{ij} + B_{ij}, \tag{1}$$

i.e.

$$\tilde{C} = \tilde{A} + \tilde{B}. \tag{2}$$

(b)

$$C_{ij} = \int u_i^* A B u_j \, d\tau. \tag{3}$$

The operator B operating on the function u_j gives (in general) another function, which can be expanded in terms of the us themselves, i.e. we may put

$$Bu_j = \sum_k d_k u_k. \tag{4}$$

The coefficient d_k is equal to the matrix element B_{kj}. We can see this by multiplying (4) by u_n^* and integrating, which gives

$$B_{nj} = \int u_n^* B u_j \, d\tau = \sum_k d_k \int u_n^* u_k \, d\tau = d_n. \tag{5}$$

In the last step we have used the orthonormal property of the us. So

$$Bu_j = \sum_k B_{kj} u_k. \tag{6}$$

From (3) and (6)

$$C_{ij} = \sum_k B_{kj} \int u_i^* A u_k \, d\tau = \sum_k A_{ik} B_{kj}, \tag{7}$$

which is the rule for matrix multiplication, i.e.

$$\tilde{C} = \tilde{A}\tilde{B}. \tag{8}$$

5.2 (a) The matrix elements are, in general, A_{ij}, where i, j take values 1, 2, 3, corresponding to $m = 1, 0, -1$ respectively. To find the elements of the matrix $\tilde{L}_x$, we first note that, from (5.1.6),

$$L_x \phi_j = \sum_k (L_x)_{kj} \phi_k. \tag{1}$$

To obtain the elements of the first column of the matrix $\tilde{L}_x$, put $j = 1$ and use the first equation in (4.2.2). The coefficients of the ϕs on the right-hand side are the elements $(L_x)_{11}$, $(L_x)_{21}$, $(L_x)_{31}$, in this case 0, $\hbar/\sqrt{2}$, 0. The coefficients of the ϕs on the right-hand of the second equation in (4.2.2), are the elements of the second column, and similarly for the third column. The matrix is given below.

To obtain the elements for the matrix $\tilde{L}_y$, we need to find the results of L_y operating in turn on ϕ_1, ϕ_0, ϕ_{-1}. They are obtained from the relation

$$L_y = \frac{1}{2\mathrm{i}}(L_+ - L_-), \tag{2}$$

and proceeding as in Problem 4.2. The results are

$$L_y\phi_1 = \mathrm{i}\frac{\hbar}{\sqrt{2}}\phi_0, \quad L_y\phi_0 = -\mathrm{i}\frac{\hbar}{\sqrt{2}}(\phi_1 - \phi_{-1}),$$

$$L_y\phi_{-1} = -\mathrm{i}\frac{\hbar}{\sqrt{2}}\phi_0. \tag{3}$$

As before the coefficients of the ϕs on the right-hand side of each equation give the elements of each column of the matrix.

Since the basis functions are eigenfunctions of the operator L_z, the matrix $\tilde{L}_z$ is diagonal, the diagonal elements being the eigenvalues of L_z.

The three matrices are therefore

$$\begin{array}{ccc} \tilde{L}_x & \tilde{L}_y & \tilde{L}_z \\ \frac{\hbar}{\sqrt{2}}\begin{bmatrix} 0 & 1 & 0 \\ 1 & 0 & 1 \\ 0 & 1 & 0 \end{bmatrix}, & \frac{\mathrm{i}\hbar}{\sqrt{2}}\begin{bmatrix} 0 & -1 & 0 \\ 1 & 0 & -1 \\ 0 & 1 & 0 \end{bmatrix}, & \hbar\begin{bmatrix} 1 & 0 & 0 \\ 0 & 0 & 0 \\ 0 & 0 & -1 \end{bmatrix}. \end{array} \tag{4}$$

The matrix for L^2 may be obtained from the relation

$$\tilde{L}^2 = \tilde{L}_x^2 + \tilde{L}_y^2 + \tilde{L}_z^2. \tag{5}$$

The matrix $\tilde{L}_x^2$ is

$$\frac{\hbar^2}{2}\begin{bmatrix} 0 & 1 & 0 \\ 1 & 0 & 1 \\ 0 & 1 & 0 \end{bmatrix}\begin{bmatrix} 0 & 1 & 0 \\ 1 & 0 & 1 \\ 0 & 1 & 0 \end{bmatrix} = \frac{\hbar^2}{2}\begin{bmatrix} 1 & 0 & 1 \\ 0 & 2 & 0 \\ 1 & 0 & 1 \end{bmatrix}. \tag{6}$$

The matrices $\tilde{L}_y^2$ and $\tilde{L}_z^2$ are obtained in the same way, and the three matrices are added to give the matrix $\tilde{L}^2$, which is

$$2\hbar^2\begin{bmatrix} 1 & 0 & 0 \\ 0 & 1 & 0 \\ 0 & 0 & 1 \end{bmatrix}. \tag{7}$$

(b) The eigenvalues λ of the matrix

$$\begin{bmatrix} 0 & 1 & 0 \\ 1 & 0 & 1 \\ 0 & 1 & 0 \end{bmatrix}$$

are given by the equation

$$\begin{vmatrix} -\lambda & 1 & 0 \\ 1 & -\lambda & 1 \\ 0 & 1 & -\lambda \end{vmatrix} = 0. \tag{8}$$

Multiplying up the determinant we have

$$\lambda(\lambda^2 - 2) = 0, \quad \text{i.e.} \quad \lambda = 0, \pm\sqrt{2}. \tag{9}$$

Multiplying these values by the factor $\hbar/\sqrt{2}$, we have $0, \pm\hbar$ for the eigenvalues of $\tilde{L}_x$.

We see from (7) that the matrix $\tilde{L}^2$ is the unit matrix multiplied by $2\hbar^2$. Thus it has three equal eigenvalues with the value $l(l+1)\hbar^2$, where $l = 1$.

Comment

If we had chosen the eigenfunctions of the operator L_x as basis, instead of those of L_z, the matrix $\tilde{L}_x$ would have been diagonal, while $\tilde{L}_y$ and $\tilde{L}_z$ would not. This reflects the fact that the operators L_x, L_y, L_z do not commute, so, apart from the special case $l = 0$, no two of these operators can have the same set of eigenfunctions, which means that we cannot have more than one of the matrices diagonal.

The operator L^2 commutes with each of the three component operators. Since matrices obey the same algebraic relations as the corresponding operators, the matrix $\tilde{L}^2$ must commute with each of the matrices $\tilde{L}_x$, $\tilde{L}_y$, $\tilde{L}_z$. Whichever linear combination of ϕ_1, ϕ_0, ϕ_{-1} we use as basis, the matrix $\tilde{L}^2$ is the same as (7). Since a unit matrix commutes with any matrix of the same order, we see that the required commutation property holds for $\tilde{L}^2$ and the other three matrices.

5.3 (a) Denote the raising and lowering operators for spin by S_+ and S_-. They satisfy (5.33) and (5.34) with $j = \frac{1}{2}$, $m = \pm\frac{1}{2}$. Substituting these values in the two equations and putting

$$\alpha = \phi_{1/2,1/2}, \qquad \beta = \phi_{1/2,-1/2}, \tag{1}$$

gives

$$S_+\alpha = 0, \qquad S_+\beta = \hbar\alpha,$$
$$S_-\alpha = \hbar\beta, \qquad S_-\beta = 0. \tag{2}$$

Now

$$S_x = \tfrac{1}{2}(S_+ + S_-), \qquad S_y = -\tfrac{1}{2}\mathrm{i}(S_+ - S_-). \tag{3}$$

Therefore

$$S_x\alpha = \tfrac{1}{2}\hbar\beta, \qquad S_x\beta = \tfrac{1}{2}\hbar\alpha,$$
$$S_y\alpha = \tfrac{1}{2}\mathrm{i}\hbar\beta, \qquad S_y\beta = -\tfrac{1}{2}\mathrm{i}\hbar\alpha. \tag{4}$$

Since α and β are eigenfunctions of S_z with eigenvalues $\pm\frac{1}{2}\hbar$, we also have

$$S_z\alpha = \tfrac{1}{2}\hbar\alpha, \quad S_z\beta = -\tfrac{1}{2}\hbar\beta. \tag{5}$$

The Pauli spin operators are defined by $S_x = \dfrac{\hbar}{2}\sigma_x$, etc. Therefore

$$\sigma_x\alpha = \beta, \quad \sigma_y\alpha = \mathrm{i}\beta, \qquad \sigma_z\alpha = \alpha,$$
$$\sigma_x\beta = \alpha, \quad \sigma_y\beta = -\mathrm{i}\alpha, \quad \sigma_z\beta = -\beta. \tag{6}$$

(b) The eigenfunctions and eigenvalues of the operators σ_x and σ_y are readily obtained from (6). Thus

$$\sigma_x(\alpha + \beta) = (\alpha + \beta), \qquad \sigma_x(\alpha - \beta) = -(\alpha - \beta),$$
$$\sigma_y(\alpha + \mathrm{i}\beta) = (\alpha + \mathrm{i}\beta), \quad \sigma_y(\alpha - \mathrm{i}\beta) = -(\alpha - \mathrm{i}\beta). \tag{7}$$

Since α and β are normalised, the eigenfunctions of σ_x and σ_y are normalised by the factor $1/\sqrt{2}$. The eigenvalues follow from inspection of (7).

(c) The matrices $\tilde{\sigma}_x$, $\tilde{\sigma}_y$, $\tilde{\sigma}_z$ are obtained from the definition in (5.11), and the relations in (6) and (5.9). Thus

$$(\sigma_x)_{\alpha\alpha} = \langle\alpha|\sigma_x\alpha\rangle = \langle\alpha|\beta\rangle = 0, \tag{8}$$
$$(\sigma_x)_{\alpha\beta} = \langle\alpha|\sigma_x\beta\rangle = \langle\alpha|\alpha\rangle = 1, \tag{9}$$

and so on. Alternatively, the matrices may be obtained from (6) by the same method as in Problem 5.2. Thus to obtain the first column of the matrix $\tilde{\sigma}_x$, we look at the coefficients of α and β on the right-hand side of the first equation of (6). They are 0 and 1. The second column is given by the coefficients of α and β on the right-hand side of the fourth equation of (6), namely, 1 and 0. Thus the matrix for $\tilde{\sigma}_x$ is

$$\begin{bmatrix} 0 & 1 \\ 1 & 0 \end{bmatrix}.$$

Similarly for $\tilde{\sigma}_y$ and $\tilde{\sigma}_z$. The matrices $\tilde{\sigma}_x$, $\tilde{\sigma}_y$, $\tilde{\sigma}_z$ are known as *Pauli spin matrices*.

5.4 (a) The spin state function ψ_P of P is the eigenfunction of S_x with eigenvalue $+\hbar/2$. So $\psi_P = (\alpha + \beta)/\sqrt{2}$.

(b) No. In each case the two beams emerging from the Stern–Gerlach apparatus would contain equal numbers of atoms.

(c) The difference between P and Q could be determined by passing each beam through a Stern–Gerlach magnet with its magnetic field along the x axis. For beam P all the atoms would emerge in the same channel, corresponding to x-component of spin angular momentum equal to $+\hbar/2$. For beam Q two beams would emerge with equal numbers of atoms. For this beam the same result would occur whatever the direction of the magnetic field of the Stern–Gerlach magnet.

5.5 To find the relative number of atoms in the two beams emerging from the Stern–Gerlach magnet with its magnetic field along the x axis, we expand the spin state function ψ in terms of the eigenfunctions of S_x. Thus

$$\begin{aligned}\psi &= (e_1\alpha + e_2\beta)/\sqrt{2} \\ &= c_1(\alpha + \beta)/\sqrt{2} + c_2(\alpha - \beta)/\sqrt{2}. \qquad (1)\end{aligned}$$

The relative number of atoms in the two emerging beams is given by $|c_1|^2/|c_2|^2$.

To obtain c_1 and c_2, we equate the coefficients of α and β in (1).

$$e_1 = c_1 + c_2, \quad e_2 = c_1 - c_2. \qquad (2)$$

So

$$c_1 = (e_1 + e_2)/2, \quad c_2 = (e_1 - e_2)/2. \qquad (3)$$

Thus

$$|c_1|^2 = \tfrac{1}{4}\{|e_1|^2 + |e_2|^2 + \langle e_1^* e_2\rangle + \langle e_2^* e_1\rangle\} = \tfrac{1}{2}, \qquad (4)$$

since

$$|e_1|^2 = |e_2|^2 = 1, \quad \langle e_1^* e_2\rangle = \langle e_2^* e_1\rangle = 0. \qquad (5)$$

(We have used the result – see the solution to Problem 4.8 – that, whenever we have the product of two *e*s with random phases between them, we must average over all values of the phase.) Similarly

$$|c_2|^2 = \tfrac{1}{2}. \qquad (6)$$

We would, of course, have obtained the same result if we had expanded ψ in terms of the eigenfunction of S_y, or indeed in terms of the eigenfunctions of a spin operator corresponding to *any* direction in space.

5.6 We first need the Pauli spin operator for the direction D. This is obtained from the classical result that, if a system has angular momentum S_z along z, and S_x along x, its angular momentum along D is

$$S_D = \cos\theta\, S_z + \sin\theta\, S_x. \tag{1}$$

The algebraic relation between quantum mechanical operators is the same as between the corresponding classical quantities. Therefore (1) gives the operator S_D for the component of angular momentum along the direction D in terms of the operators S_z and S_x. The Pauli spin operator for the direction D is therefore

$$\sigma_D = \cos\theta\, \sigma_z + \sin\theta\, \sigma_x. \tag{2}$$

After emerging from the first Stern–Gerlach apparatus the state of the atoms is the eigenfunction of σ_D with eigenvalue +1. We therefore require the eigenfunctions of σ_D. From (2) and (5.3.6)

$$\sigma_D\alpha = \cos\theta\, \alpha + \sin\theta\, \beta, \tag{3}$$

$$\sigma_D\beta = \sin\theta\, \alpha - \cos\theta\, \beta. \tag{4}$$

If $(a\alpha + b\beta)$ is an eigenfunction of σ_D with eigenvalue λ,

$$\sigma_D(a\alpha + b\beta) = \lambda(a\alpha + b\beta). \tag{5}$$

From (3) and (4)

$$\sigma_D(a\alpha + b\beta) = (a\cos\theta + b\sin\theta)\alpha + (a\sin\theta - b\cos\theta)\beta. \tag{6}$$

Equating the coefficients of α and β on the right-hand sides of (5) and (6) gives

$$a\cos\theta + b\sin\theta = \lambda a, \quad a\sin\theta - b\cos\theta = \lambda b, \tag{7}$$

whence

$$\frac{a}{b} = \frac{\sin\theta}{\lambda - \cos\theta} = \frac{\lambda + \cos\theta}{\sin\theta}. \tag{8}$$

These equations give $\lambda = \pm 1$, as we expect, because the eigenvalues of σ_D must be the same as those of σ_x, σ_y, σ_z.

When a beam of atoms in the state $\psi = a\alpha + b\beta$ is passed through a Stern–Gerlach apparatus with its magnetic field along the z axis, the relative numbers of emerging atoms in the states α and β is a^2/b^2. For $\lambda = 1$, (8) gives

$$\frac{a^2}{b^2} = \left(\frac{\sin\theta}{1-\cos\theta}\right)^2 = \frac{\cos^2\frac{1}{2}\theta}{\sin^2\frac{1}{2}\theta}. \tag{9}$$

If we wish to do the algebra in terms of matrices rather than operators we use (2) to obtain the matrix $\tilde{\sigma}_D$. Thus

$$\tilde{\sigma}_D = \cos\theta\,\tilde{\sigma}_z + \sin\theta\,\tilde{\sigma}_x$$

$$= \begin{bmatrix} \cos\theta & \sin\theta \\ \sin\theta & -\cos\theta \end{bmatrix}. \tag{10}$$

The eigenfunction equation is

$$\begin{bmatrix} \cos\theta & \sin\theta \\ \sin\theta & -\cos\theta \end{bmatrix}\begin{bmatrix} a \\ b \end{bmatrix} = \lambda\begin{bmatrix} a \\ b \end{bmatrix}. \tag{11}$$

Multiplication then gives the same equations as (7), and the rest of the calculation is unchanged. This is a demonstration of the essential similarity of the operator and matrix methods. They are alternative ways of doing the 'book-keeping' and always lead to the same answers.

5.7 (a) We first note that if S_{1-}, S_{2-}, S_- are the lowering operators for the spin angular momenta of electrons 1 and 2, and for the resultant angular momentum, then

$$S_- = S_{1-} + S_{2-}. \tag{1}$$

This relation follows from the definition of a lowering operator, and the fact that each component of a resultant vector is the sum of the corresponding components of the constituent vectors. Thus

$$S_{1-} + S_{2-} = S_{1x} - \mathrm{i}S_{1y} + S_{2x} - \mathrm{i}S_{2y} = S_x - \mathrm{i}S_y = S_-. \tag{2}$$

Now consider the four product functions

$$\alpha\alpha,\ \alpha\beta,\ \beta\alpha,\ \beta\beta. \tag{3}$$

If m_1, m_2 are the eigenvalues of the operators S_{1z}, S_{2z}, then the state $\alpha\alpha$ corresponds to $m_1 = \frac{1}{2}$, $m_2 = \frac{1}{2}$. It is an eigenstate of the operator S_z with eigenvalue $M = m_1 + m_2 = 1$. The state $\alpha\beta$ corresponds to $m_1 = \frac{1}{2}$, $m_2 = -\frac{1}{2}$, $M = 0$. Similarly for the other two. Thus the four states in (3) have

$$M = 1, 0, 0, -1. \tag{4}$$

The two electrons have spin quantum numbers $s_1 = s_2 = \frac{1}{2}$. Therefore, from (5.24), the quantum number for the operator S^2 has the values

$$S = 1, 0. \tag{5}$$

Each eigenfunction Φ_{SM} of the operators S^2 and S_z can be represented as a linear combination of the product functions in (3) that have the specified value of M.

Consider the function Φ_{11}. It has eigenvalue $M = 1$, and since there is only one product function in (3) with this eigenvalue, Φ_{11} must be equal to it. Thus

$$\Phi_{11} = \alpha\alpha. \tag{6}$$

To obtain the expression for Φ_{10} we use (1), applying S_- to the left-hand side of (6), and $S_{1-} + S_{2-}$ to the right-hand side, i.e.

$$S_-\Phi_{11} = (S_{1-} + S_{2-})\alpha\alpha. \tag{7}$$

For a general angular momentum eigenfunction ϕ_{jm}, we have – see (5.34) –

$$J_-\phi_{jm} = \{j(j+1) - m(m-1)\}^{1/2}\hbar\phi_{j,m-1}. \tag{8}$$

With $j = S = 1$, and $m = M = 1$, (8) becomes

$$S_-\Phi_{11} = \surd 2\hbar\Phi_{10}. \tag{9}$$

We now apply $S_{1-} + S_{2-}$ to the right-hand side of (6). S_{1-} operates only on the first α, which relates to electron 1; it treats the second α as a constant. Similarly, S_{2-} operates only on the second α, treating the first α as a constant. The state α corresponds to $j = \frac{1}{2}$, $m = \frac{1}{2}$, and the state β to $j = \frac{1}{2}$, $m = -\frac{1}{2}$. Inserting these values in (8) gives

$$(S_{1-} + S_{2-})\alpha\alpha = \hbar(\beta\alpha + \alpha\beta). \tag{10}$$

From (7), (9), and (10) we have the required result

$$\Phi_{10} = (\alpha\beta + \beta\alpha)/\surd 2. \tag{11}$$

We could proceed in the same way to obtain Φ_{1-1}, i.e. we could apply the operator S_- to the left-hand side of (11) and $S_{1-} + S_{2-}$ to the right-hand side. But of course it is much quicker to note that there is only one product with the eigenvalue $M = -1$, namely $\beta\beta$. So

$$\Phi_{1-1} = \beta\beta. \tag{12}$$

Finally we require Φ_{00}. Since $M = 0$, we know that Φ_{00} is a linear combination of $\alpha\beta$ and $\beta\alpha$. Let $p\alpha\beta + q\beta\alpha$ be the required function. Since $S_-\Phi_{00} = 0$, we have

$$(S_{1-} + S_{2-})(p\alpha\beta + q\beta\alpha) = \hbar(p\beta\beta + q\beta\beta) = 0. \tag{13}$$

Therefore

$$p + q = 0. \tag{14}$$

Thus

$$\Phi_{00} = (\alpha\beta - \beta\alpha)/\sqrt{2}. \tag{15}$$

An alternative, and simpler, method, is to make use of the fact that all the Φs are orthogonal to each other. The functions Φ_{10} and Φ_{00} are both linear combinations of $\alpha\beta$ and $\beta\alpha$. To make a function orthogonal to the right-hand side of (11), we exchange the coefficients, which in this case are equal, and insert a minus sign, obtaining the result in (15).

(b) Denote the lowering operators for the resultant, orbital, and spin angular momenta by J_-, L_-, S_-. From (8)

$$L_-\phi_1 = \sqrt{2}\hbar\phi_0, \quad L_-\phi_0 = \sqrt{2}\hbar\phi_{-1}, \quad L_-\phi_{-1} = 0,$$

$$S_-\alpha = \hbar\beta, \quad S_-\beta = 0. \tag{16}$$

For $l = 1$, $s = \frac{1}{2}$, the possible values of j are $\frac{3}{2}$, $\frac{1}{2}$.

We start with the Φ for the maximum values of j, m_j, namely $j = m_j = \frac{3}{2}$. The function $\Phi_{3/2,3/2}$ must be represented by the single product eigenfunction with $m_l + m_s = \frac{3}{2}$. So

$$\Phi_{3/2,3/2} = \phi_1\alpha. \tag{17}$$

To obtain $\Phi_{3/2,1/2}$, we use the relation $J_- = L_- + S_-$ and evaluate

$$J_-\Phi_{3/2,3/2} = (L_- + S_-)\phi_1\alpha. \tag{18}$$

From (8)

$$J_-\Phi_{3/2,3/2} = \sqrt{3}\hbar\Phi_{3/2,1/2}. \tag{19}$$

On the right-hand side of (18), L_- operates only on ϕ_1 and treats α as a constant. Similarly, S_- operates only on α and treats ϕ_1 as a constant. Thus, from (16),

$$(L_- + S_-)\phi_1\alpha = \hbar(\sqrt{2}\phi_0\alpha + \phi_1\beta), \tag{20}$$

whence

$$\Phi_{3/2,1/2} = (\sqrt{2}\phi_0\alpha + \phi_1\beta)/\sqrt{3}. \tag{21}$$

We obtain $\Phi_{3/2,-1/2}$ by the same procedure.

$$J_-\Phi_{3/2,1/2} = (L_- + S_-)(\sqrt{2}\phi_0\alpha + \phi_1\beta)/\sqrt{3}, \tag{22}$$

which gives

$$\Phi_{3/2,-1/2} = (\phi_{-1}\alpha + \sqrt{2}\phi_0\beta)/\sqrt{3}. \tag{23}$$

Clearly

$$\Phi_{3/2,-3/2} = \phi_{-1}\beta. \tag{24}$$

The eigenfunctions $\Phi_{1/2,1/2}$ and $\Phi_{1/2,-1/2}$ are most easily obtained from the orthogonality condition. $\Phi_{1/2,1/2}$ is orthogonal to $\Phi_{3/2,1/2}$. Thus

$$\Phi_{1/2,1/2} = (\phi_0\alpha - \sqrt{2}\phi_1\beta)/\sqrt{3}. \tag{25}$$

Similarly $\Phi_{1/2,1/2}$ is orthogonal to $\Phi_{3/2,-1/2}$. Thus

$$\Phi_{1/2,-1/2} = (\sqrt{2}\phi_{-1}\alpha - \phi_0\beta)/\sqrt{3}. \tag{26}$$

The results of this problem are summarised in Table 5.1.

Table 5.1. *Expansion of the eigenfunctions Φ in terms of the product eigenfunctions for the two cases in Problem 5.7.*

$s_1 = s_2 = \frac{1}{2}$		
M_S	$S = 1$	$S = 0$
1	$\alpha\alpha$	
0	$(\alpha\beta + \beta\alpha)/\sqrt{2}$	$(\alpha\beta - \beta\alpha)/\sqrt{2}$
−1	$\beta\beta$	
$l = 1, s = \frac{1}{2}$		
m_j	$j = \frac{3}{2}$	$j = \frac{1}{2}$
3/2	$\phi_1\alpha$	
1/2	$(\sqrt{2}\phi_0\alpha + \phi_1\beta)/\sqrt{3}$	$(\phi_0\alpha - \sqrt{2}\phi_1\beta)/\sqrt{3}$
−1/2	$(\phi_{-1}\alpha + \sqrt{2}\phi_0\beta)/\sqrt{3}$	$(\sqrt{2}\phi_{-1}\alpha - \phi_0\beta)/\sqrt{3}$
−3/2	$\phi_{-1}\beta$	

Comment

It is stated on p. 76 that a spin state function may be regarded as a vector in Hilbert space. Such a representation is not restricted to spin functions. Any set of N orthonormal eigenfunctions of a Hermitian operator may be regarded as a set of unit vectors in an N-dimensional Hilbert space. Moreover, we may use functions which are products of space and spin eigenfunctions as unit vectors. For example, the set of six orthonormal functions,

$$\phi_1\alpha,\ \phi_0\alpha,\ \phi_{-1}\alpha,\ \phi_1\beta,\ \phi_0\beta,\ \phi_{-1}\beta, \tag{27}$$

may be taken as unit vectors in a six-dimensional Hilbert space. They are said to span the space, meaning that any vector in the space represents a linear combination of the six unit vectors. The set of functions in (27) is not unique for spanning this space. Other orthonormal sets of six functions, each set being a linear combination of the states in (27), may be used as unit vectors. The set of Φs in the second half of Table 5.1 is an example of another such set.

5.8 When a magnetic field **B** is applied along the z axis to an atom, the Hamiltonian has the form

$$H = H^0 + \frac{\mu_B B}{\hbar}(L_z + 2S_z) + \frac{2W}{\hbar^2}\mathbf{L}\cdot\mathbf{S}. \tag{1}$$

The notation and the physical origin of the terms is explained in Problem 5.10. The constant W is of the order of the spin–orbit energy E_0. (The solution to Problem 5.10 shows that $E_0 = 3W$ in the present case.) The force, and hence the deflection, of an atom in a Stern–Gerlach apparatus with its magnetic field gradient along the z axis is proportional to the z-component of the magnetic dipole moment of the atom. If B is small compared to E_0/μ_B, the Hamiltonian is effectively of the form $H^0 + (2W/\hbar^2)\mathbf{L}\cdot\mathbf{S}$, and the eigenfunctions are the Φ_{jm_j} functions – see p. 80. The beam entering the first Stern–Gerlach magnet has $j = \frac{3}{2}$. The magnetic dipole moment is proportional to m_j. So four beams emerge from the first magnet, corresponding to the values $m_j = \frac{3}{2}, \frac{1}{2}, -\frac{1}{2}, -\frac{3}{2}$. Their states are represented by the functions $\Phi_{3/2,3/2}$, $\Phi_{3/2,1/2}$, $\Phi_{3/2,-1/2}$, $\Phi_{3/2,-3/2}$.

If $B \gg E_0/\mu_B$, we may neglect the term in $\mathbf{L}\cdot\mathbf{S}$ in the Hamiltonian, which becomes

$$H = H^0 + \frac{\mu_B B}{\hbar}(L_z + 2S_z). \tag{2}$$

The eigenfunctions are the product functions of Problem 5.7(b), namely $\phi_{m_l}\alpha$ and $\phi_{m_l}\beta$, where $m_l = 1, 0, -1$. For each of the four Stern–Gerlach magnets, the atoms entering are in a state Φ_{jm_j}, and the atoms in each beam that emerges are in a state given by one of the product functions. We use the results of Problem 5.7(b) to express Φ_{jm_j} in terms of the product functions. The number of emerging beams is equal to the number of terms in the expansion, and the relative number of atoms in each beam is given by the square of the coefficient of the term. For the beam entering with $m_j = \frac{3}{2}$, we have

$$\Phi_{3/2,3/2} = \phi_1 \alpha. \tag{3}$$

So this beam is not split. For the beam $m_j = \frac{1}{2}$, we have

$$\Phi_{3/2,1/2} = (\sqrt{2}\phi_0\alpha + \phi_1\beta)\sqrt{3}. \tag{4}$$

So this beam is split into two, with relative number of atoms 2:1. Similarly the beam $m_j = -\frac{1}{2}$ is split into two, with relative number 1:2, and the beam $m_j = -\frac{3}{2}$ is not split.

The results are summarised in Fig. 5.1. The numbers in bold are the numbers of atoms in each beam expressed as a fraction of the number entering the first weak-B magnet. Note that for a given Φ_{jm_j}, each $\phi_{m_l}\phi_{m_s}$ in the expansion has the same value of $m_l + m_s = m_j$, but a different value of $m_l + 2m_s$, and hence experiences a different deflecting force in the strong-B magnet.

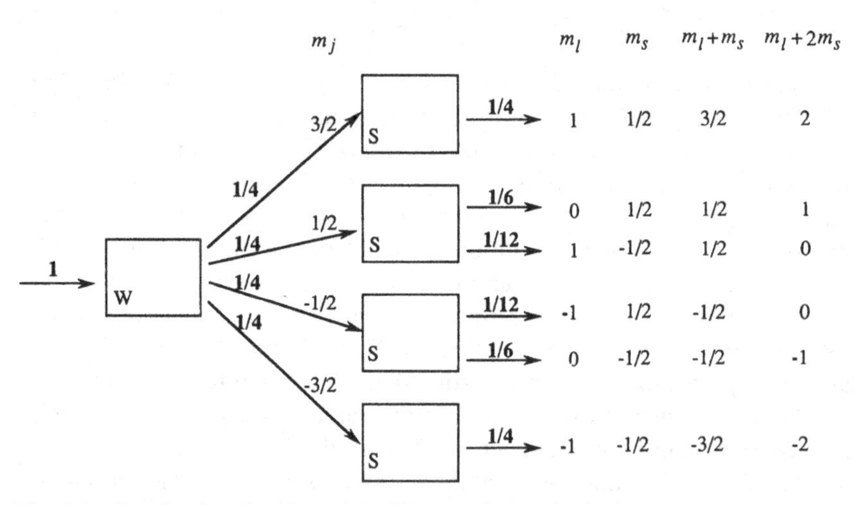

Fig. 5.1. *Results for Problem 5.8. W is a Stern–Gerlach magnet with a weak magnetic field, and S is one with a strong magnetic field. A number in bold gives the number of atoms in the beam as a fraction of the number entering the weak-B Stern–Gerlach magnet.*

5.9 (a) When $B = 0$, $H = W\boldsymbol{\sigma}_e \cdot \boldsymbol{\sigma}_p$. Use the results in (5.3.6), remembering that an e (electron) operator operates only on the first α or β, and a p (proton) operator operates only on the second α or β. Then

$$\begin{aligned}\boldsymbol{\sigma}_e \cdot \boldsymbol{\sigma}_p \alpha\alpha &= (\sigma_{ex}\sigma_{px} + \sigma_{ey}\sigma_{py} + \sigma_{ez}\sigma_{pz})\alpha\alpha \\ &= \beta\beta - \beta\beta + \alpha\alpha = \alpha\alpha. \end{aligned} \tag{1}$$

Similarly

$$\boldsymbol{\sigma}_e\cdot\boldsymbol{\sigma}_p\alpha\beta = 2\beta\alpha - \alpha\beta, \tag{2}$$

$$\boldsymbol{\sigma}_e\cdot\boldsymbol{\sigma}_p\beta\alpha = 2\alpha\beta - \beta\alpha, \tag{3}$$

$$\boldsymbol{\sigma}_e\cdot\boldsymbol{\sigma}_p\beta\beta = \beta\beta, \tag{4}$$

whence

$$\boldsymbol{\sigma}_e\cdot\boldsymbol{\sigma}_p(\alpha\beta + \beta\alpha) = \alpha\beta + \beta\alpha, \tag{5}$$

$$\boldsymbol{\sigma}_e\cdot\boldsymbol{\sigma}_p(\alpha\beta - \beta\alpha) = -3(\alpha\beta - \beta\alpha). \tag{6}$$

So the states $\alpha\alpha$, $(\alpha\beta + \beta\alpha)\surd 2$, $\beta\beta$ are normalised eigenfunctions of $W\boldsymbol{\sigma}_e\cdot\boldsymbol{\sigma}_p$ with eigenvalue W, and the state $(\alpha\beta - \beta\alpha)\surd 2$ is a normalised eigenfunction with eigenvalue $-3W$.

Alternatively we may use the fact that the operator $\boldsymbol{\sigma}_e\cdot\boldsymbol{\sigma}_p$ commutes with $(\boldsymbol{\sigma}_e + \boldsymbol{\sigma}_p)^2$ – see (5.25) – and therefore, from the results of Problem 5.7(a), its eigenfunctions are $\alpha\alpha$, $(\alpha\beta + \beta\alpha)\surd 2$, $\beta\beta$, $(\alpha\beta - \beta\alpha)\surd 2$. It only remains to find the eigenvalues. We have

$$\boldsymbol{\sigma}_e\cdot\boldsymbol{\sigma}_p = \tfrac{1}{2}\{(\boldsymbol{\sigma}_e + \boldsymbol{\sigma}_p)^2 - \sigma_e^2 - \sigma_p^2\}. \tag{7}$$

Then, if Φ is any one of the four eigenfunctions,

$$\boldsymbol{\sigma}_e\cdot\boldsymbol{\sigma}_p\Phi = \tfrac{4}{2}\{S(S+1) - \tfrac{3}{4} - \tfrac{3}{4}\}\Phi, \tag{8}$$

where $S = 1$ for the first three eigenfunctions, and $S = 0$ for the last one. (The factor 4 arises because a Pauli spin operator is $2/\hbar$ times the corresponding angular momentum operator. So, in units of $\hbar$, the eigenvalue of the square of a Pauli operator is 4 times the eigenvalue of the square of the angular momentum operator.) Eq. (8) becomes

$$\boldsymbol{\sigma}_e\cdot\boldsymbol{\sigma}_p\Phi = \Phi, \qquad S = 1, \tag{9}$$

$$\boldsymbol{\sigma}_e\cdot\boldsymbol{\sigma}_p\Phi = -3\Phi, \quad S = 0, \tag{10}$$

which gives the same two eigenvalues as before.

(b) When B is not zero, the Hamiltonian is $H = \varepsilon\sigma_{ez} + W\boldsymbol{\sigma}_e\cdot\boldsymbol{\sigma}_p$. We have

$$\sigma_{ez}\alpha\alpha = \alpha\alpha, \qquad \sigma_{ez}\alpha\beta = \alpha\beta,$$

$$\sigma_{ez}\beta\alpha = -\beta\alpha, \quad \sigma_{ez}\beta\beta = -\beta\beta. \tag{11}$$

From these results and those of (1) to (4)

$$H\alpha\alpha = (\varepsilon + W)\alpha\alpha, \tag{12}$$

$$H\alpha\beta = (\varepsilon - W)\alpha\beta + 2W\beta\alpha, \tag{13}$$

$$H\beta\alpha = 2W\alpha\beta - (\varepsilon + W)\beta\alpha, \tag{14}$$

$$H\beta\beta = (-\varepsilon + W)\beta\beta. \tag{15}$$

Eqs. (12) and (15) show that $\alpha\alpha$ and $\beta\beta$ remain eigenstates. Their respective eigenvalues are $W + \varepsilon$, $W - \varepsilon$.

Eqs. (13) and (14) show that the matrix of H on the basis of the functions $\alpha\beta$ and $\beta\alpha$ is

$$\begin{bmatrix} \varepsilon - W & 2W \\ 2W & -(\varepsilon + W) \end{bmatrix}.$$

It is readily shown that $p\alpha\beta + q\beta\alpha$ is an eigenvector of this matrix with eigenvalue λ, where

$$\frac{q}{p} = \frac{\lambda + W - \varepsilon}{2W} = \frac{2W}{\lambda + W + \varepsilon}, \tag{16}$$

whence

$$\lambda = -W \pm (4W^2 + \varepsilon^2)^{1/2}, \quad \text{and} \quad \frac{q}{p} = \frac{-\varepsilon \pm (4W^2 + \varepsilon^2)^{1/2}}{2W}. \tag{17}$$

For $B = 0$, i.e. $\varepsilon = 0$, $\lambda = W, -3W$ as before, and $q/p = \pm 1$, giving the same eigenfunctions as before.

(c) The energy values

$$W + \varepsilon,\ (4W^2 + \varepsilon^2)^{1/2} - W,\ W - \varepsilon,\ -(4W^2 + \varepsilon^2)^{1/2} - W$$

are shown schematically as functions of B in Fig. 5.2. For $\varepsilon = \mu_B B \ll W$, the energies are, to first order in ε/W,

$$W + \varepsilon,\ W,\ W - \varepsilon,\ -3W, \tag{18}$$

i.e. the triply degenerate eigenvalue W is split into three, one staying constant and the other two varying linearly with B, with equal and opposite slopes. The other eigenvalue $-3W$ remains constant. For $\mu_B B \gg W$, the energies tend to $\varepsilon + W$, $\varepsilon - W$, $-\varepsilon + W$, $-\varepsilon - W$.

The angular momentum information comes from the eigenstates. We have seen that, for $B = 0$, the eigenstates are the functions Φ_{11}, Φ_{10}, Φ_{1-1}, Φ_{00} of Problem 5.7(a). For the lines a and c in Fig. 5.2, the respective eigenstates are $\Phi_{11} = \alpha\alpha$ and $\Phi_{1-1} = \beta\beta$, for all values of B. For curve b, the eigenstate at $B = 0$ is Φ_{10}, and for $\mu_B B \gg W$ it is $\alpha\beta$, or, expressed more accurately, it tends to $\alpha\beta$ as B tends to infinity. At intermediate values of B, the eigenstate can be expressed either as a linear combination of the states Φ_{10} and Φ_{00}, or as a linear combination of

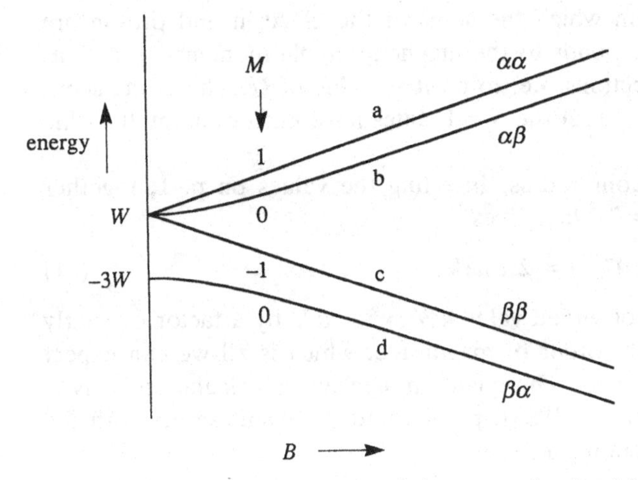

Fig. 5.2. *The hyperfine structure of the energy levels of hydrogen in a magnetic field* **B**. *As $B \to 0$, the eigenfunctions of the Hamiltonian tend to Φ_{SM}; the three states tending to the energy W have $S = 1$, and the state tending to $-3W$ has $S = 0$; the M values are as shown. As $B \to \infty$, the eigenfunctions tend to the product eigenfunctions of σ_{ez}, σ_{pz}; the first function in the product shown refers to the electron, and the second to the proton.*

the states $\alpha\beta$ and $\beta\alpha$. Similarly, for curve d, the eigenstate at $B = 0$ is Φ_{00}, and for $\mu_B B \gg W$ it is $\beta\alpha$. Again, at intermediate values of B, it can be expressed either as a linear combination of the states Φ_{10} and Φ_{00}, or as a linear combination of the states $\alpha\beta$ and $\beta\alpha$.

(d) For two magnetic dipoles with moments $\boldsymbol{\mu}_e$ and $\boldsymbol{\mu}_p$, a distance r apart, and aligned as shown in Fig. 5.3, the potential energy given by classical electromagnetism is

$$U = -\frac{\mu_0}{2\pi}\frac{\mu_e\mu_p}{r^3}. \tag{19}$$

So the change in energy when one of the dipoles is reversed is

$$\Delta E = 2U = \frac{\mu_0}{\pi}\frac{\mu_e\mu_p}{r^3}. \tag{20}$$

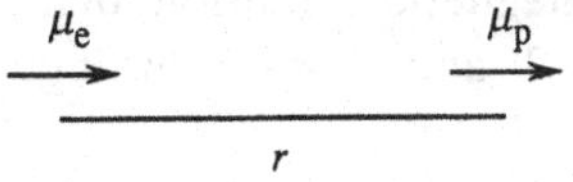

μ_e μ_p

r

Fig. 5.3. *Arrangement of magnetic dipoles of the electron and proton.*

Note that, the state in which the spins of the electron and proton are aligned ($S = 1$), corresponds to the magnetic dipole moments $\boldsymbol{\mu}_e$ and $\boldsymbol{\mu}_p$ being in opposite directions, i.e. to positive value of U. This is in accord with the previous result that the $S = 1$ states have higher energy than the $S = 0$ state.

Take $r = a_0$, the Bohr radius. Inserting the values on p. 1, together with $\mu_e = \mu_B$, and $\mu_p = 2.79\mu_N$, gives

$$\Delta E = 3.5 \times 10^{-25}\,\mathrm{J} = 2.2\,\mu\mathrm{eV}. \tag{21}$$

This differs from the observed value $4W = 5.9\,\mu\mathrm{eV}$ by a factor of nearly 3, but it is of the right order of magnitude, which is all we can expect from a classical calculation. The quantum mechanical calculation is given in Bransden and Joachain (1983), p. 241, and the result agrees with the measured value to about 0.1%.

Comments

(1) The results for $\mu_B B \gg W$ have a simple physical interpretation. If we take W to be zero, the energy values correspond to the independent alignment of the magnetic dipole moments of the electron and proton either parallel or antiparallel to the direction of the magnetic field **B**, which is along the z axis. The energy is positive when the magnetic dipole moment is opposed to **B**. The electron state α corresponds to the spin angular momentum vector along $+z$. For the negatively charged electron, the magnetic dipole moment is then in the direction of $-z$. So the states a, b in Fig. 5.2, which correspond to the electron state α, have higher energy than the states c, d. The same reasoning applies to the alignment of the spin angular momentum of the proton, but, since the proton has positive charge, its magnetic dipole moment is in the same direction as its angular momentum vector. So, for the proton, the state β has a higher energy than α. The reason this is not true for the curves a and b is the neglect of the term $B\mu_p\sigma_{pz}$ in the original Hamiltonian. If this term were included, then, for very large values of B, the β state would have the higher energy. For the same electron state, the separation of the energies of the two states with opposite proton spins is always small compared to the separation of the states with opposite electron spins, because the magnetic dipole moment of the proton is about 1/660 that of the electron.

(2) The value of $4W$, the hyperfine separation, is obtained experimentally by measuring the frequency of the electromagnetic radiation that brings about transitions between the states $S = 1$ and $S = 0$. Putting $4W = h\nu$ gives

$$\nu = 1420\,\mathrm{MHz}, \tag{22}$$

which is in the microwave region. The value of ν has been determined to a very high degree of precision – a few parts in 10^{11}. The wavelength corresponding to this frequency is 21 cm, and measurement of the intensity of this radiation from regions in outer space provides information on the distribution of hydrogen in the universe.

(3) The affect of a magnetic field on the frequency of spectral lines is known as the *Zeeman effect* after its discoverer, who originally observed it in sodium. The Zeeman effect in the 21 cm line of hydrogen is used by astronomers to detect the magnetic field in the interstellar clouds of hydrogen in our Galaxy. The magnetic field is extremely small, being about 10^{-10} to 10^{-9} T. The difference ν_{diff} in the frequencies of the transitions from the levels $S = 1$, $M = \pm 1$ to the level $S = 0$, $M = 0$ is only a few hertz for such magnetic fields. The spread in frequency of each line, mainly due to the Doppler effect of the moving atoms, is of the order of 10^4 Hz. Nevertheless, ν_{diff} can still be measured – at least approximately – because the two lines are circularly polarised in opposite directions, which enables the detecting apparatus to determine ν_{diff} directly.

5.10 (a)
$$\begin{aligned}\mathbf{L}\cdot\mathbf{S} &= L_xS_x + L_yS_y + L_zS_z \\ &= \tfrac{1}{2}\{(L_x + \mathrm{i}L_y)(S_x - \mathrm{i}S_y) + (L_x - \mathrm{i}L_y)(S_x + \mathrm{i}S_y)\} + L_zS_z \\ &= \tfrac{1}{2}(L_+S_- + L_-S_+) + L_zS_z. \end{aligned} \tag{1}$$

(b) From (1) the Hamiltonian is

$$H = \frac{\varepsilon}{\hbar}(L_z + 2S_z) + \frac{W}{\hbar^2}(L_+S_- + L_-S_+ + 2L_zS_z). \tag{2}$$

For the operators L_z and S_z, we have

$$\begin{aligned} &L_z\phi_1 = \hbar\phi_1, \qquad L_z\phi_0 = 0\phi_0, \qquad L_z\phi_{-1} = -\hbar\phi_{-1}, \\ &S_z\alpha = \tfrac{1}{2}\hbar\alpha, \qquad S_z\beta = -\tfrac{1}{2}\hbar\beta, \end{aligned} \tag{3}$$

and we use (4.2.1) and (5.3.2) for the results of the application of the L_+, L_-, S_+, and S_- operators.

To obtain $H\chi_1 = H\phi_1\alpha$, we calculate the result of each term in H operating on $\phi_1\alpha$. Thus

$$\frac{1}{\hbar}(L_z + 2S_z)\phi_1\alpha = 2\phi_1\alpha, \tag{4}$$

$$\frac{1}{\hbar^2}(L_+S_- + L_-S_+)\phi_1\alpha = 0. \tag{5}$$

This result follows because $L_+\phi_1 = 0$, $S_+\alpha = 0$. Finally

$$\frac{1}{\hbar^2}L_zS_z\phi_1\alpha = \tfrac{1}{2}\phi_1\alpha. \tag{6}$$

From (2) to (6)

$$H\chi_1 = (2\varepsilon + W)\phi_1\alpha, \tag{7}$$

i.e.

$$H\chi_1 = c_{11}\chi_1, \quad \text{where} \quad c_{11} = 2\varepsilon + W. \tag{8}$$

For $\chi_2 = \phi_1\beta$, we have

$$\frac{1}{\hbar}(L_z + 2S_z)\phi_1\beta = (1 - 1)\phi_1\beta = 0, \tag{9}$$

$$\frac{1}{\hbar^2}(L_+S_- + L_-S_+)\phi_1\beta = \text{zero} + \sqrt{2}\phi_0\alpha, \tag{10}$$

$$\frac{1}{\hbar^2}L_zS_z\phi_1\beta = -\tfrac{1}{2}\phi_1\beta. \tag{11}$$

Thus

$$H\chi_2 = -W\phi_1\beta + \sqrt{2}W\phi_0\alpha = c_{22}\chi_2 + c_{23}\chi_3, \tag{12}$$

where $c_{22} = -W$, and $c_{23} = \sqrt{2}W$. The rest of the results follow in a similar way.

(c) The cs are the matrix of the operator H on the basis of the χs. From the results of (b) the matrix is

$$\left[\begin{array}{c|cc|cc|c}
2\varepsilon + W & 0 & 0 & 0 & 0 & 0 \\
\hline
0 & -W & \sqrt{2}W & 0 & 0 & 0 \\
0 & \sqrt{2}W & \varepsilon & 0 & 0 & 0 \\
\hline
0 & 0 & 0 & -\varepsilon & \sqrt{2}W & 0 \\
0 & 0 & 0 & \sqrt{2}W & -W & 0 \\
\hline
0 & 0 & 0 & 0 & 0 & -2\varepsilon + W
\end{array}\right]. \tag{13}$$

The matrix factorises into two single-element matrices at the top left- and bottom right-hand corners, together with two 2×2 matrices. These are indicated by the dashed lines in (13). The two single-element matrices have eigenvalues $W + 2\varepsilon$, and $W - 2\varepsilon$, corresponding to respective eigenfunctions χ_1 and χ_6. For the matrix of the elements c_{22}, c_{23}, c_{32}, c_{33}, the eigenvalues λ are given by

$$\begin{vmatrix} -W-\lambda & \sqrt{2}W \\ \sqrt{2}W & \varepsilon-\lambda \end{vmatrix} = 0. \tag{14}$$

Multiplying up the determinant we have

$$(\lambda + W)(\lambda - \varepsilon) = 2W^2, \tag{15}$$

with solution

$$\lambda = \tfrac{1}{2}[\varepsilon - W \pm \{(\varepsilon + W)^2 + 8W^2\}^{1/2}]. \tag{16}$$

Similarly, the matrix of the elements c_{44}, c_{45}, c_{54}, c_{55} has eigenvalues λ, given by the roots of the quadratic equation

$$(\lambda + W)(\lambda + \varepsilon) = 2W^2, \tag{17}$$

i.e.

$$\lambda = \tfrac{1}{2}[-(\varepsilon + W) \pm \{(\varepsilon - W)^2 + 8W^2\}^{1/2}]. \tag{18}$$

(d) (i) For $\varepsilon \ll W$,

$$\{(\varepsilon + W)^2 + 8W^2\}^{1/2} \approx (9W^2 + 2\varepsilon W)^{1/2} \approx 3W + \frac{\varepsilon}{3}, \tag{19}$$

$$\{(\varepsilon - W)^2 + 8W^2\}^{1/2} \approx (9W^2 - 2\varepsilon W)^{1/2} \approx 3W - \frac{\varepsilon}{3}. \tag{20}$$

So the values of E tend to

$$W \pm 2\varepsilon, \quad W \pm \frac{2\varepsilon}{3}, \quad -2W \pm \frac{\varepsilon}{3}. \tag{21}$$

For $\varepsilon \ll W$, the expression given by the vector model for the energy due to the magnetic field $\mathbf{B}$ is

$$E_{\text{mag}} = g m_j \mu_B B, \tag{22}$$

where the quantity g, known as the *Landé g factor*, is given by

$$g = 1 + \frac{j(j+1) + s(s+1) - l(l+1)}{2j(j+1)}. \tag{23}$$

See Haken and Wolf (1987), p. 204 for a discussion of the vector model.

For $B = 0$, the four states with energy W in (21) correspond to $j = \frac{3}{2}$. Inserting this value, together with $l = 1$, and $s = \frac{1}{2}$ into (23) gives $g = \frac{4}{3}$. The values of m_j in (22) are $\frac{3}{2}, \frac{1}{2}, -\frac{1}{2}, -\frac{3}{2}$. For these four values, the values of E_{mag} in (22) are

$$2\varepsilon, \quad \frac{2\varepsilon}{3}, \quad -\frac{2\varepsilon}{3}, \quad -2\varepsilon, \tag{24}$$

which agrees with the variation of E with B for the first four terms of (21). For $B = 0$, the two states with energy $-2W$ in (21) correspond to $j = \frac{1}{2}$, giving $g = \frac{2}{3}$. This value, together with $m_j = \frac{1}{2}, -\frac{1}{2}$, gives $E_{\text{mag}} = \varepsilon/3$, $-\varepsilon/3$, which agrees with (21).

(ii) For $\varepsilon \gg W$,

$$\{(\varepsilon + W)^2 + 8W^2\}^{1/2} \approx \varepsilon + W, \tag{25}$$

$$\{(\varepsilon - W)^2 + 8W^2\}^{1/2} \approx \varepsilon - W. \tag{26}$$

Thus the values of E tend to

$$2\varepsilon + W, \quad \varepsilon, \quad -W, \quad -W, \quad -\varepsilon, \quad -2\varepsilon + W. \tag{27}$$

In the vector model, for $\varepsilon \gg W$, the magnetic dipole moments due to orbital and spin angular momentum precess independently around the magnetic field vector $\mathbf{B}$. This gives rise to magnetic energy

$$E_{\text{mag}} = (m_l + 2m_s)\mu_{\text{B}} B. \tag{28}$$

The m_l, m_s values of the states with the six energies in (27) are

$$1, \tfrac{1}{2}, \quad 0, \tfrac{1}{2}, \quad -1, \tfrac{1}{2}, \quad 1, -\tfrac{1}{2}, \quad 0, -\tfrac{1}{2}, \quad -1, -\tfrac{1}{2}. \tag{29}$$

For these combinations, the values of E_{mag} are

$$2\varepsilon, \quad \varepsilon, \quad 0, \quad 0, \quad -\varepsilon, \quad -2\varepsilon, \tag{30}$$

which agree with the terms in ε in (27).

Comments

(1) Clearly the analysis of this problem is very similar to that of Problem 5.9. The present problem involves more algebra, because we are dealing with the angular momentum quantum numbers 1 and $\frac{1}{2}$. There are six product eigenfunctions (or six Φ eigenfunctions), which means a 6×6 matrix for H. In the previous problem the quantum numbers were $\frac{1}{2}$ and $\frac{1}{2}$, with four eigenfunctions, and a 4×4 matrix for H.

(2) The six energy values in the present problem are shown as functions of B in Fig. 5.4. For all values of B, the eigenfunction of H for the line a in the figure is $\Phi_{3/2,3/2} = \phi_1\alpha$, and for the line f it is $\Phi_{3/2,-3/2} = \phi_{-1}\beta$. The eigenfunctions for the other four curves are linear combinations of the χ product functions (or of the Φ functions). For example, the eigenfunction for curve b has the form

$$\psi_{\text{b}} = a_2\phi_1\beta + a_3\phi_0\alpha. \tag{31}$$

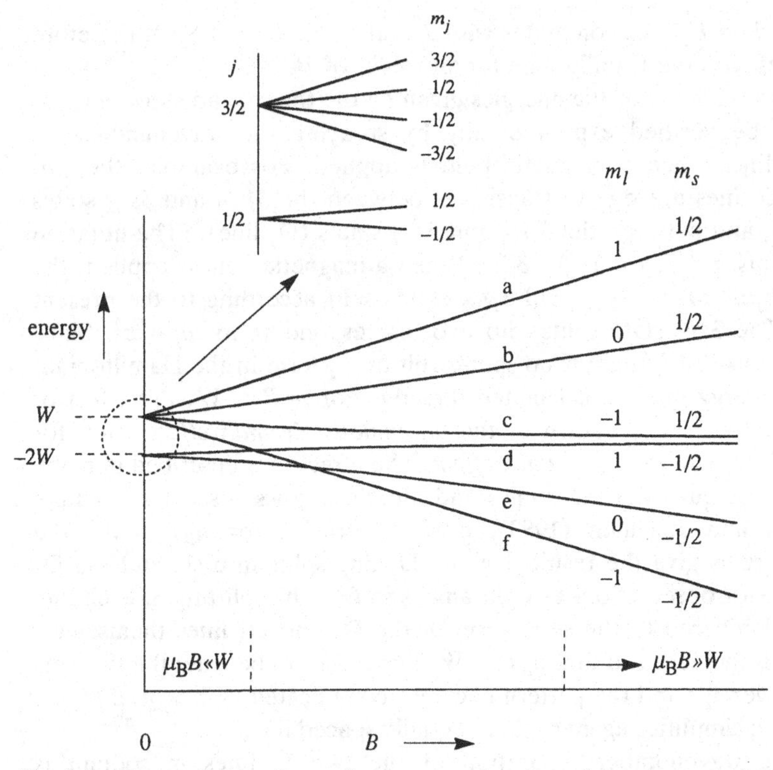

Fig. 5.4. *Energy levels of a p electron as a function of the magnetic field B. For $\mu_B B \ll W$, the eigenfunctions of the Hamiltonian are the states Φ_{jm_j} with the quantum numbers shown in the enlarged diagram at the top. For $\mu_B B \gg W$, the eigenfunctions are the product states χ ($\phi_{m_l}\alpha$ or $\phi_{m_l}\beta$) with the quantum numbers shown on the right.*

The coefficients a_2 and a_3 are functions of B. For $B = 0$

$$a_2 = \sqrt{\tfrac{1}{3}}, \quad a_3 = \sqrt{\tfrac{2}{3}}, \tag{32}$$

so ψ_b is the function $\Phi_{3/2,1/2}$ (result of Problem 5.7(b)). As B is increased from zero, a_3 rises and a_2 falls. For $\mu_B B \gg W$, a_3 tends to unity, and a_2 to zero, so that the eigenfunction for the curve b becomes the state $\phi_0\alpha$. For a fixed value of B, the eigenfunction for curve d is orthogonal to that of b, since they are eigenvectors of the matrix in (13). Thus

$$\psi_d = a_3\phi_1\beta - a_2\phi_0\alpha. \tag{33}$$

For $B = 0$, ψ_d is the function $\Phi_{1/2,1/2,}$ and for $\mu_B B \gg W$, ψ_d tends to $\phi_1\beta$. Similar statements can be made for curves c and e.

Note that, for each of the six curves, the quantum number $m_j = m_l + m_s$, has the same value for all values of B. This is because the operator $J_z = L_z + S_z$ commutes with J^2, and with L_z and S_z. It therefore commutes with the Hamiltonian for all values of B.

(3) The variation of the energies given by the theory and shown in Fig. 5.4 may be verified experimentally by studying the wavenumbers of spectral lines when a magnetic field is applied. For example, the two sodium D lines are due to transitions between the $3p_{3/2}$ and $3s_{1/2}$ states (D_2 line), and between the $3p_{1/2}$ and $3s_{1/2}$ states (D_1 line). (The notation $3p_{3/2}$ means $n = 3$, $l = 1$, $j = 3/2$.) When a magnetic field is applied, the energy levels of the $3p_{3/2}$ and $3p_{1/2}$ states split according to the present theory. The $3s_{1/2}$ state splits into two, corresponding to $m_j = \pm\frac{1}{2}$. Since $l = 0$ for this state, there is no spin–orbit or L_z term in the Hamiltonian, and the energy may be calculated directly. For $\mu_B B \ll W$, the effect of the magnetic field is known as the *anomalous Zeeman effect*, and, for $\mu_B B \gg W$, as the *Paschen–Back effect*. The effects are described in many textbooks of quantum mechanics and atomic physics – see for example Bransden and Joachain (1983), p. 207. Briefly, for $\mu_B B \ll W$, the selection rules give the result that the D_2 line splits into six and the D_1 into four (for observations at right angles to **B**). The splitting of each line is small compared to the separation of the D_1 and D_2 lines themselves. When B is increased so that $\mu_B B \sim W$, this ceases to be true; the two sets of lines overlap, and the pattern becomes complicated. When $\mu_B B \gg W$, the pattern simplifies again to three equally spaced lines.

(4) The wavenumber separation of the two D lines in sodium is $\tilde{\nu} = 1720\,\mathrm{m}^{-1}$. Converting this to an energy and equating it to $\mu_B B_0$, we have

$$B_0 = \frac{ch}{\mu_B}\tilde{\nu} = 37\,\mathrm{T}, \tag{34}$$

which gives the order of magnitude of the magnetic field dividing the anomalous Zeeman and Paschen–Back regions. The value in (34) is a large magnetic field. So for sodium it is usually the anomalous Zeeman effect that is observed. However, for lithium, the wavenumber separation of the $2p_{3/2}$ and the $2p_{1/2}$ states is $34\,\mathrm{m}^{-1}$, giving $B_0 = 0.73\,\mathrm{T}$. It is therefore relatively easy to observe the anomalous Zeeman, intermediate, and Paschen–Back behaviour in this element.

6

Approximation methods. Time-independent perturbation theory, variational method

Summary of theory

1 Time-independent perturbation theory

The method is used when the Hamiltonian of a system has the form

$$H = H^{(0)} + H^{(1)}, \tag{1}$$

where $H^{(1)} \ll H^{(0)}$, and the eigenfunctions u_j and eigenvalues E_j of $H^{(0)}$, given by

$$H^{(0)}u_j = E_j u_j, \tag{2}$$

are known for all j. Let ϕ be an eigenfunction of H with eigenvalue E, i.e.

$$H\phi = E\phi. \tag{3}$$

The method is to express E and ϕ as a series, thus

$$E = E^{(0)} + E^{(1)} + E^{(2)} + \ldots \tag{4}$$

$$\phi = \phi^{(0)} + \phi^{(1)} + \phi^{(2)} + \ldots \tag{5}$$

In these expressions, $E^{(0)}$ is E_n, one of the E_j values, and $\phi^{(0)}$ is the corresponding u_n. (Assume for the moment that E_n is non-degenerate.) The other terms with superscript (r), are proportional to the rth power of matrix elements of the operator $H^{(1)}$ on the basis of the states u_j, and are known as *rth-order* terms. We shall be concerned only with the terms $E^{(1)}$, $\phi^{(1)}$, and $E^{(2)}$.

The theory gives the following results.

$$E^{(1)} = \langle n|H^{(1)}|n\rangle, \tag{6}$$

$$\phi^{(1)} = \sum_{j\neq n} c_j u_j, \tag{7}$$

where

$$c_j = \frac{\langle j|H^{(1)}|n\rangle}{E_n - E_j}, \tag{8}$$

$$E^{(2)} = \sum_{j \neq n} \frac{|\langle j|H^{(1)}|n\rangle|^2}{E_n - E_j}. \tag{9}$$

If the unperturbed energy E_n is degenerate, we cannot use these results as they stand, because in (8) and (9) the denominators are zero for some of the terms in the summation, and the quantities become infinite. The method then proceeds as follows. Suppose the unperturbed energy E_n is g-fold degenerate, i.e. the states $u_{n1}, u_{n2}, \ldots, u_{ng}$ have the same energy E_n. We find g orthogonal, linear combinations of these states – call them v_{np}, where p takes the values 1 to g – such that

$$\langle v_{np}|H^{(1)}|v_{nq}\rangle = 0, \quad \text{for } p \neq q. \tag{10}$$

Then

$$E_p^{(1)} = \langle v_{np}|H^{(1)}|v_{np}\rangle, \quad p = 1 \text{ to } g. \tag{11}$$

The value of $E_p^{(1)}$ may be different for some or all values of p, i.e. the perturbation Hamiltonian may partially or fully remove the degeneracy in the unperturbed energy. Note that the states v_{np}, like the u_{np}, are eigenfunctions of $H^{(0)}$ with the same eigenvalue E_n. The unperturbed function $\phi^{(0)}$ for a particular p value is v_{np}. The first-order function $\phi^{(1)}$ is given by the same expression as (7) and (8), except that the state $|n\rangle$ in the matrix element is replaced by $|v_{np}\rangle$, and, in the summation over j, all the g degenerate states are excluded.

2 Variational method

The method is used to obtain an upper bound for E_0, the energy of the ground state of a system whose Hamiltonian H is known, but the eigenfunctions and eigenvalues of H are not known. The method is to guess a plausible state function ψ to represent the eigenfunction of the ground state. Then for the function ψ, the expectation value of the energy is

$$\langle H\rangle = \langle \psi|H|\psi\rangle. \tag{12}$$

The theory shows that $\langle H\rangle \geqslant E_0$, the equality holding only if ψ happens to be equal to the ground-state wave function. If ψ contains a variable parameter β, then $\langle H\rangle$ depends on β. The variational method is to vary β, and find the minimum value of $\langle H\rangle$, which is again greater or equal to E_0. If ψ is a function of several parameters $\beta_1, \ldots, \beta_g$, each β is varied in turn, and the overall minimum is again an upper bound for E_0. If there are sufficient variable parameters, the minimum is expected to be a close approximation to E_0. The method can be extended to obtain an upper bound for the energy of a higher level, but it is then more complicated, as it is necessary to make the trial function orthogonal to the eigenfunctions of all the lower states.

Problems

6.1 The potential function of a one-dimensional oscillator of mass m, and angular frequency ω, is

$$V(x) = \tfrac{1}{2}kx^2 + cx^4,$$

where the second term is small compared to the first.

(a) Show that, to first order, the effect of the anharmonic term is to change the energy E_0 of the ground state by $3c(\hbar/2m\omega)^2$.

(b) What would be the first-order effect of an additional term in x^3 in the potential?

6.2 A particle of mass m moves in a one-dimensional simple harmonic potential $V^{(0)} = \frac{1}{2}kx^2$, with angular frequency $\omega = \sqrt{(k/m)}$. A small perturbing term $V^{(1)} = \frac{1}{2}\delta k\, x^2$ is added to $V^{(0)}$.

(a) Show that the first- and second-order perturbations to the energy of the ground state are

$$E^{(1)} = \tfrac{1}{4}\frac{\delta k}{k}\hbar\omega, \qquad E^{(2)} = -\tfrac{1}{16}\left(\frac{\delta k}{k}\right)^2 \hbar\omega.$$

(b) How do these expressions relate to the exact expression for the energy?

6.3 A particle of mass m and charge e oscillates in a one-dimensional harmonic potential with angular frequency ω.

(a) Show, using perturbation theory, that the effect of an applied electric field $\mathcal{E}$ is to lower all the energy levels by $e^2\mathcal{E}^2/2m\omega^2$.

(b) Compare this with the classical result.

6.4 Use first-order perturbation theory to show that, for the atom considered in Problem 5.10, the variation of energy with magnetic field B is 2ε, $\frac{2}{3}\varepsilon$, $-\frac{2}{3}\varepsilon$, -2ε for the $j=\frac{3}{2}$ state, and $\frac{1}{3}\varepsilon$, $-\frac{1}{3}\varepsilon$ for the $j=\frac{1}{2}$ state, where $\varepsilon = \mu_B B \ll W$.

6.5 If spin effects are neglected, the four states of the hydrogen atom with $n=2$ have the same energy E^0. Show that, when an electric field $\mathcal{E}$ is applied to hydrogen atoms in these states, the resulting first-order energies are

$$E^0 \pm 3a_0 e\mathcal{E}, \quad E^0, \quad E^0.$$

6.6 Suppose that in the last problem the states u_{210} and u_{200} are not degenerate, but have energies $E^0 + \Delta$ and $E^0 - \Delta$.

(a) Show that the matrix of $H^{(0)} + H^{(1)}$ on the basis of these states has eigenvalues

$$E^0 \pm \sqrt{(\Delta^2 + \varepsilon^2)},$$

where $\varepsilon = 3a_0 e\mathcal{E}$.

(b) Compare the limiting values of these eigenvalues for (i) $\varepsilon \gg \Delta$, and (ii) $\varepsilon \ll \Delta$ with the results of perturbation theory.

(c) Sketch the form of $E^0 \pm (\varepsilon^2 + \Delta^2)^{1/2}$ as a function of $\mathcal{E}$, labelling the curves with as much information as possible about the eigenvectors.

(d) If the wavenumber separation of the states u_{210} and u_{200} is $36\ \mathrm{m}^{-1}$, calculate the value of the electric field for $\varepsilon = \Delta$.

6.7 The effect of the finite size of the nucleus is to raise the energies of the electronic states from the theoretical values based on a point nucleus.

(a) Show from first-order perturbation theory that, if the proton is regarded (for simplicity) as a thin uniform spherical shell of charge of radius b, the fractional change in the energy of the ground state of the hydrogen atom is $4b^2/3a_0^2$.

(b) Why is the fractional energy change so much greater for a μ^- meson orbiting a lead nucleus?

6.8 (a) Taking a trial wave function proportional to $\exp(-\beta r)$, where β is a variable parameter, use the variational method to obtain an upper limit for the energy of the ground state of the hydrogen atom in terms of atomic constants.

(b) Comment on your result.

6.9 It was shown in Problem 3.3 that for an attractive one-dimensional square-well potential there is at least one bound state. Use this result and the variational principle to prove that there is at least one bound state for a one-dimensional attractive potential of *any* shape.

Solutions

6.1 (a) From (6.6)

$$E^{(1)} = \langle n|H^{(1)}|n\rangle = c\langle 0|x^4|0\rangle. \tag{1}$$

Use the annihilation and creation operators. From (4.17)

$$x = \left(\frac{\hbar}{2m\omega}\right)^{1/2}(a + a^\dagger). \tag{2}$$

Therefore

$$\langle 0|x^4|0\rangle = \left(\frac{\hbar}{2m\omega}\right)^2\langle 0|(a + a^\dagger)(a + a^\dagger)(a + a^\dagger)(a + a^\dagger)|0\rangle. \tag{3}$$

If the operator A is the product of a number of as and $a^\dagger$s in some arbitrary order, then

$$\langle n|A|n\rangle = 0, \tag{4}$$

unless the number of as and the number of $a^\dagger$s are equal. For if they are not, $A|n\rangle$ is a state different from $|n\rangle$, and hence orthogonal to $|n\rangle$. Furthermore, a product operator like $a^\dagger aa^\dagger a$ also gives zero when $n = 0$, because $a|0\rangle = 0$. Therefore, the only non-zero contributions to $\langle 0|x^4|0\rangle$ come from the two terms $aaa^\dagger a^\dagger$ and $aa^\dagger aa^\dagger$. From the relations

$$a^\dagger|n\rangle = \sqrt{(n+1)}|n+1\rangle, \quad a|n\rangle = \sqrt{n}|n-1\rangle, \tag{5}$$

it is readily verified that

$$\langle 0|aaa^\dagger a^\dagger|0\rangle = 2, \quad \langle 0|aa^\dagger aa^\dagger|0\rangle = 1. \tag{6}$$

From (1), (3), and (6)

$$E^{(1)} = 3c\left(\frac{\hbar}{2m\omega}\right)^2. \tag{7}$$

(b) The first-order effect of a term in x^3 in the potential is zero. This is true for any odd power of x, because the operator A above is then the product of an odd number of as and $a^\dagger$s, so the number of as and $a^\dagger$s cannot be equal.

6.2 (a) The Hamiltonian is $H = H^{(0)} + H^{(1)}$, with $H^{(1)} = \frac{1}{2}\delta k\, x^2$. The basis states are the eigenfunctions $|n\rangle$ of $H^{(0)}$, the Hamiltonian for the unperturbed simple harmonic oscillator. These functions are orthogonal and normalised, i.e.

$$\langle n|n'\rangle = \delta_{nn'}. \tag{1}$$

From (6.6), the first-order energy term is

$$E^{(1)} = \langle n|H^{(1)}|n\rangle = \tfrac{1}{2}\delta k\langle 0|x^2|0\rangle, \tag{2}$$

since the unperturbed state is the ground state $n = 0$. From (4.9.13)

$$\langle 0|x^2|0\rangle = \frac{\hbar}{2m\omega}. \tag{3}$$

Thus

$$E^{(1)} = \tfrac{1}{2}\delta k\frac{\hbar}{2m\omega} = \tfrac{1}{4}\frac{\delta k}{k}\hbar\omega. \tag{4}$$

The second-order energy term is

$$E^{(2)} = \sum_{j\neq 0}\frac{|H^{(1)}_{j0}|^2}{E_0 - E_j}, \tag{5}$$

$$H^{(1)}_{j0} = \tfrac{1}{2}\delta k\langle j|x^2|0\rangle = \tfrac{1}{2}\delta k\frac{\hbar}{2m\omega}\langle j|(a + a^\dagger)(a + a^\dagger)|0\rangle. \tag{6}$$

Since j cannot be zero, the ladder properties of the operators a, $a^\dagger$ show that only $j = 2$ gives a non-zero $H^{(1)}_{j0}$. The pair of operators that converts $|0\rangle$ into $|2\rangle$ is $a^\dagger a^\dagger$. From the relation

$$a^\dagger|n\rangle = \sqrt{(n+1)}|n+1\rangle, \tag{7}$$

we have

$$a^\dagger a^\dagger|0\rangle = \sqrt{2}|2\rangle. \tag{8}$$

Thus

$$H^{(1)}_{20} = \frac{\sqrt{2}}{2}\delta k\frac{\hbar}{2m\omega}. \tag{9}$$

Further

$$E_0 = \tfrac{1}{2}\hbar\omega, \quad E_2 = \tfrac{5}{2}\hbar\omega. \tag{10}$$

Inserting these values into (5) gives

$$E^{(2)} = -\tfrac{1}{16}(\delta k)^2\frac{\hbar}{m^2\omega^3} = -\tfrac{1}{16}\left(\frac{\delta k}{k}\right)^2\hbar\omega. \tag{11}$$

(b) We do not need perturbation theory to solve this problem. The calculation can be done exactly – and more simply. The ground-state energy of the unperturbed oscillator is

$$E_0 = \tfrac{1}{2}\hbar\omega = \tfrac{1}{2}\hbar\left(\frac{k}{m}\right)^{1/2}. \tag{12}$$

So the energy of the perturbed oscillator is

$$E = \tfrac{1}{2}\hbar\left(\frac{k+\delta k}{m}\right)^{1/2}$$

$$= \tfrac{1}{2}\hbar\omega\left(1+\frac{\delta k}{k}\right)^{1/2}$$

$$= \tfrac{1}{2}\hbar\omega\left\{1+\tfrac{1}{2}\frac{\delta k}{k}-\tfrac{1}{8}\left(\frac{\delta k}{k}\right)^2+0\left(\frac{\delta k}{k}\right)^3\right\}. \qquad (13)$$

The binomial expansion in the last line is valid only when $\delta k \ll k$, i.e. when $H^{(1)} \ll H^{(0)}$.

We see that the terms in $\delta k/k$ and $(\delta k/k)^2$ in (13) correspond respectively to the 1st and 2nd order energies in perturbation theory. Normally we are not able to obtain an exact solution of the problem, which is why we resort to perturbation theory. The present calculation shows that, when $\delta k \ll k$, perturbation theory provides a good approximation to the correct energy.

6.3 (a) The potential due to the applied electric field $\mathscr{E}$ is

$$H^{(1)} = -e\mathscr{E}x. \qquad (1)$$

Let the unperturbed state be $|n\rangle$, an eigenfunction of the Hamiltonian $H^{(0)}$ for a simple harmonic oscillator. From (6.6), the first-order change in the energy is

$$E^{(1)} = -e\mathscr{E}\langle n|x|n\rangle = 0, \qquad (2)$$

from the reasoning of Problem 6.1(b).

We therefore look at the second-order correction. From (6.9) this is

$$E^{(2)} = \sum_{j\neq n}\frac{|H^{(1)}_{jn}|^2}{E_n - E_j}, \qquad (3)$$

where

$$H^{(1)}_{jn} = -e\mathscr{E}\langle j|x|n\rangle. \qquad (4)$$

As in the previous two problems we express x in terms of the operators a and $a^\dagger$ by means of (4.17), and use the orthogonal properties of the eigenfunctions $|n\rangle$. The matrix element $\langle j|x|n\rangle$ is zero unless $x|n\rangle$ contains a term in $|j\rangle$. It therefore follows that the sum over j in (3) contains only two terms, namely $j = n \pm 1$. From (6.1.2) and (6.1.5)

$$x|n\rangle = (\hbar/2m\omega)^{1/2}\{\sqrt{n}|n-1\rangle + \sqrt{(n+1)}|n+1\rangle\}. \qquad (5)$$

So

$$\langle n+1|x|n\rangle = (\hbar/2m\omega)^{1/2}\surd(n+1),$$
$$\langle n-1|x|n\rangle = (\hbar/2m\omega)^{1/2}\surd n. \tag{6}$$

Further

$$E_n - E_{n+1} = -\hbar\omega, \quad E_n - E_{n-1} = \hbar\omega. \tag{7}$$

From the results of (3) to (7)

$$E^{(2)} = e^2\mathcal{E}^2 \frac{\hbar}{2m\omega}\frac{1}{\hbar\omega}\{-(n+1)+n\} = -\frac{e^2\mathcal{E}^2}{2m\omega^2}. \tag{8}$$

In fact the present problem, like the last one, can be solved exactly in quantum mechanics without recourse to perturbation theory, and the calculation shows that the expression in (8) is the correct value for the energy change.

(b) Classically the potential function is

$$V = \tfrac{1}{2}m\omega^2x^2 - \mathcal{E}ex, \tag{9}$$

which has a minimum when

$$\frac{\mathrm{d}V}{\mathrm{d}x} = m\omega^2 x - \mathcal{E}e = 0, \quad \text{i.e. when} \quad x = \frac{\mathcal{E}e}{m\omega^2}. \tag{10}$$

Thus for $\mathcal{E} = 0$, the equilibrium position is $x = 0$ with energy $E = V = 0$, and for $\mathcal{E} = \mathcal{E}$ the equilibrium position is $x = \mathcal{E}e/m\omega^2$. Substituting this value of x in (9) gives

$$E = V_{\min} = -\frac{e^2\mathcal{E}^2}{2m\omega^2}. \tag{11}$$

It can be seen that the classical and quantum results are the same, which is not surprising as the quantum result does not contain $\hbar$.

6.4 Denote the Hamiltonian by $H^{(0)} + H^{(1)}$, where $H^{(0)}$ represents the non-magnetic terms plus $(2W/\hbar^2)\mathbf{L}\cdot\mathbf{S}$, and

$$H^{(1)} = \frac{\varepsilon}{\hbar}(L_z + 2S_z). \tag{1}$$

A set of eigenfunctions of the operator $H^{(0)}$, corresponding to $j = \frac{3}{2}$, are the four functions listed in Table 5.1, p. 96. They are

$$\Phi_{3/2,3/2} = \phi_1\alpha, \tag{2}$$
$$\Phi_{3/2,1/2} = \surd\tfrac{2}{3}\phi_0\alpha + \surd\tfrac{1}{3}\phi_1\beta, \tag{3}$$

$$\Phi_{3/2,-1/2} = \sqrt{\tfrac{1}{3}}\phi_{-1}\alpha + \sqrt{\tfrac{2}{3}}\phi_0\beta, \tag{4}$$

$$\Phi_{3/2,-3/2} = \phi_{-1}\beta. \tag{5}$$

The functions have the same eigenvalue for $H^{(0)}$. We therefore need to apply the procedure of degenerate perturbation theory, i.e. we have to choose linear combinations of the four Φs – denote them by v_g, $g = 1$ to 4 – such that, for each g, $H^{(1)}v_g$ is orthogonal to $v_{g'}$ $(g' \neq g)$. Then

$$E_g^{(1)} = \langle g|H^{(1)}|g\rangle. \tag{6}$$

We first calculate $H^{(1)}\Phi_{j,m_j}$, using the results

$$L_z\phi_m = m\hbar\phi_m, \quad S_z\alpha = \tfrac{1}{2}\hbar\alpha, \quad S_z\beta = -\tfrac{1}{2}\hbar\beta. \tag{7}$$

Thus

$$H^{(1)}\Phi_{3/2,3/2} = (\phi_1\alpha + \phi_1\alpha)\varepsilon = (2\phi_1\alpha)\varepsilon, \tag{8}$$

$$H^{(1)}\Phi_{3/2,1/2} = (\sqrt{\tfrac{1}{3}}\phi_1\beta + \sqrt{\tfrac{2}{3}}\phi_0\alpha - \sqrt{\tfrac{1}{3}}\phi_1\beta)\varepsilon = (\sqrt{\tfrac{2}{3}}\phi_0\alpha)\varepsilon, \tag{9}$$

$$H^{(1)}\Phi_{3/2,-1/2} = -(\sqrt{\tfrac{2}{3}}\phi_0\beta)\varepsilon, \tag{10}$$

$$H^{(1)}\Phi_{3/2,-3/2} = (-2\phi_{-1}\beta)\varepsilon. \tag{11}$$

It is clear that the four Φs satisfy the required condition as they stand, i.e. they are the four vs. We could have deduced this without calculating $H^{(1)}\Phi_{j,m_j}$, because the operator $(L_z + 2S_z)$ commutes with J_z. Therefore, when $H^{(1)}$ operates on Φ_{j,m_j} it gives an eigenfunction of J_z with the same m_j value, which is orthogonal to one with a different m_j.

However, we still need the results of (8) to (11) to obtain the first-order energy values. We use the result that the functions $\phi_m\alpha$ and $\phi_m\beta$ are orthogonal and normalised. Then

$$m_j = 3/2, \quad E^{(1)} = \langle\phi_1\alpha|H^{(1)}|\phi_1\alpha\rangle = 2\varepsilon\langle\phi_1\alpha|\phi_1\alpha\rangle = 2\varepsilon, \tag{12}$$

$$m_j = 1/2, \quad E^{(1)} = \varepsilon\langle\sqrt{\tfrac{2}{3}}\phi_0\alpha + \sqrt{\tfrac{1}{3}}\phi_1\beta|\sqrt{\tfrac{2}{3}}\phi_0\alpha\rangle = \tfrac{2}{3}\varepsilon, \tag{13}$$

$$m_j = -1/2, \quad E^{(1)} = -\varepsilon\langle\sqrt{\tfrac{1}{3}}\phi_{-1}\alpha + \sqrt{\tfrac{2}{3}}\phi_0\beta|\sqrt{\tfrac{2}{3}}\phi_0\beta\rangle = -\tfrac{2}{3}\varepsilon, \tag{14}$$

$$m_j = -3/2, \quad E^{(1)} = -2\varepsilon\langle\phi_{-1}\beta|\phi_{-1}\beta\rangle = -2\varepsilon. \tag{15}$$

The first-order energies for the two $j = \tfrac{1}{2}$ states are obtained in the same way. The two eigenfunctions of $H^{(0)}$ are given in Table 5.1. They are

$$\Phi_{1/2,1/2} = \sqrt{\tfrac{1}{3}}\phi_0\alpha - \sqrt{\tfrac{2}{3}}\phi_1\beta, \tag{16}$$

$$\Phi_{1/2,-1/2} = \sqrt{\tfrac{2}{3}}\phi_{-1}\alpha - \sqrt{\tfrac{1}{3}}\phi_0\beta. \tag{17}$$

Then

$$H^{(1)}\Phi_{1/2,1/2} = (\sqrt{\tfrac{1}{3}}\phi_0\alpha)\varepsilon, \tag{18}$$

$$H^{(1)}\Phi_{1/2,-1/2} = -(\sqrt{\tfrac{1}{3}}\phi_0\beta)\varepsilon, \tag{19}$$

giving

$$m_j = 1/2, \qquad E^{(1)} = \tfrac{1}{3}\varepsilon, \tag{20}$$

$$m_j = -1/2, \quad E^{(1)} = -\tfrac{1}{3}\varepsilon. \tag{21}$$

The results (12) to (15) and (20), (21) agree with those in Problem 5.10(d) for $\varepsilon \ll W$ – see (5.10.21).

6.5 The Hamiltonian $H^{(1)}$ for an electron with charge $-e$ in an electric field $\mathcal{E}$ along the z axis is $H^{(1)} = e\mathcal{E}z$. Therefore, the first-order energy change for the state u_{nlm} in the hydrogen atom is

$$E^{(1)} = e\mathcal{E}\int u^*_{nlm} z u_{nlm}\, \mathrm{d}\tau = 0. \tag{1}$$

The result follows because the state u_{nlm} has a definite parity, which is positive for even l and negative for odd l. Since z has odd parity, the integrand in (1) has odd parity, whatever the parity of u_{nlm}, and the integral over all space of a function with negative parity is zero. We can obtain a non-zero value for $E^{(1)}$ only from a state of mixed parity, i.e. from a linear combination of states with opposite parities.

Denote the four states of the problem by u_{211}, u_{21-1}, u_{210}, u_{200}. Numbering them from 1 to 4, we first evaluate the matrix elements $H^{(1)}_{ij}$ on the basis of these functions. Since the four us have definite parity, the diagonal elements $H^{(1)}_{ii} = 0$. Moreover, z commutes with L_z, so both u_{nlm} and zu_{nlm} are eigenfunctions of L_z with the same eigenvalue $m\hbar$. Therefore, since the u functions with different values of m are orthogonal, it follows that the only non-zero off-diagonal matrix elements are those corresponding to functions with the same value of m, namely, u_{210} and u_{200}. To evaluate them we need the explicit forms of the functions given on pp. 52 and 54. Then, putting $z = r\cos\theta$, we have

$$\begin{aligned} H^{(1)}_{34} = H^{(1)}_{43} &= e\mathcal{E}\int u^*_{210} z u_{200}\, \mathrm{d}\tau \\ &= e\mathcal{E}\left(\frac{1}{2a_0}\right)^4 \int_0^\pi \cos^2\theta \sin\theta\, \mathrm{d}\theta \\ &\qquad \times \int_0^\infty r^4\left(2 - \frac{r}{a_0}\right)\exp\left(-\frac{r}{a_0}\right)\mathrm{d}r \\ &= -3e\mathcal{E}a_0. \end{aligned} \tag{2}$$

In evaluating the integral in r we have used the result

$$\int_0^\infty r^n \exp\left(-\frac{r}{a_0}\right) \mathrm{d}r = n!a_0^{n+1}. \tag{3}$$

The matrix of $H^{(1)}$ on the basis of the four u states is thus

$$\begin{bmatrix} 0 & 0 & 0 & 0 \\ 0 & 0 & 0 & 0 \\ 0 & 0 & 0 & -\varepsilon \\ 0 & 0 & -\varepsilon & 0 \end{bmatrix}, \tag{4}$$

where $\varepsilon = 3e\mathscr{E}a_0$.

Since the four unperturbed states have the same energy, we apply degenerate perturbation theory. The eigenvalues of the matrix $H^{(1)}$ are the first-order energy corrections, and its eigenvectors are the linear combinations of the us that give the correct unperturbed wave functions. It is clear from the form of the matrix in (4) that two of the eigenvalues are zero, with corresponding eigenvectors u_{211} and u_{21-1}. We need consider only the part of the matrix that relates to u_{210} and u_{200}, i.e.

$$\begin{bmatrix} 0 & -\varepsilon \\ -\varepsilon & 0 \end{bmatrix}. \tag{5}$$

It is readily shown that the eigenvalues λ of this matrix are

$$\lambda = \varepsilon, \qquad \text{with eigenfunction } (u_{210} - u_{200})/\surd 2, \tag{6}$$

and

$$\lambda = -\varepsilon, \quad \text{with eigenfunction } (u_{210} + \mathrm{u}_{200})/\surd 2. \tag{7}$$

The function u_{210} has negative parity, while u_{200} has positive parity, so the two eigenfunctions have mixed parity.

To first order the energies are thus E^0, E^0, $E^0 + 3e\mathscr{E}a_0$, $E^0 - 3e\mathscr{E}a_0$.

Comments

(1) The effect on the energy levels of an atom is known as the *Stark effect*. We see from the above reasoning that there is a first-order effect only when there are degenerate states with different values of l. The degeneracy in hydrogen occurs because the electrostatic potential for the electron, due to the nuclear charge, has a simple $1/r$ dependence. However, in alkali atoms, the presence of the inner electrons changes the potential of the valency electron from the $1/r$ form for states with low l values, where the valency electron penetrates the inner shells. Hence the first-order Stark effect does not occur for these states.

(2) The first-order Stark effect for the $n = 2$ states in hydrogen means that for these states the atom behaves as though it has a permanent

electric dipole moment of magnitude $3ea_0$, which can be oriented either parallel, antiparallel, or at right-angles to an external electric field. In general, the ground states of atoms and of nuclei are non-degenerate, so it follows that they do not possess a permanent electric dipole moment.

The second-order Stark effect, which we have not discussed, occurs in general in all states. It gives an energy term proportional to $\mathscr{E}^2$, which corresponds to an *induced* electric dipole moment.

6.6 (a) The matrix of $H^{(0)} + H^{(1)}$ on the basis of u_{210} and u_{200} is

$$\begin{bmatrix} E^0 + \Delta & -\varepsilon \\ -\varepsilon & E^0 - \Delta \end{bmatrix}.$$

The $H^{(0)}$ operator gives the diagonal elements. The results of the last problem give the off-diagonal elements, which come from the $H^{(1)}$ operator. We find the eigenvalues and eigenfunctions of the matrix in the usual way. If $pu_{210} + qu_{200}$ is an eigenfunction with eigenvalue λ, then

$$\frac{q}{p} = \frac{\lambda - E^0 - \Delta}{-\varepsilon} = \frac{-\varepsilon}{\lambda - E^0 + \Delta}, \tag{1}$$

whence

$$\lambda = E^0 \pm \surd(\Delta^2 + \varepsilon^2).$$

(b) (i) For $\varepsilon \gg \Delta$, we have

$$(\varepsilon^2 + \Delta^2)^{1/2} = \varepsilon\left(1 + \frac{\Delta^2}{\varepsilon^2}\right)^{1/2} \approx \varepsilon + \frac{\Delta^2}{2\varepsilon}. \tag{2}$$

So the energy values are approximately $E^0 \pm \varepsilon \pm (\Delta^2/2\varepsilon)$. The first two terms are the same as those obtained in the last problem by degenerate perturbation theory. This is to be expected, since for $\varepsilon \gg \Delta$ the initial energy separation of the two states is small compared to the perturbation energy. The effect of a small non-zero Δ is shown by the term $\Delta^2/2\varepsilon$.

(ii) For $\varepsilon \ll \Delta$, we have

$$(\varepsilon^2 + \Delta^2)^{1/2} = \Delta\left(1 + \frac{\varepsilon^2}{\Delta^2}\right)^{1/2} \approx \Delta + \frac{\varepsilon^2}{2\Delta}. \tag{3}$$

The energy values are approximately $E^0 \pm \Delta \pm (\varepsilon^2/2\Delta)$. In this situation, when the initial energy separation of the two states is large compared to the perturbation energy, there is effectively no degeneracy, and hence no linear Stark effect. The quadratic term $\pm(\varepsilon^2/2\Delta)$ is the same as that given by second-order perturbation theory. In the expression for $E^{(2)}$ in (6.9),

$\langle j|H^{(1)}|n\rangle = -\varepsilon$, while $E_n - E_j$ is 2Δ for the initial state u_{210}, and -2Δ for the initial state u_{200}.

(c) The expression $E^0 \pm (\varepsilon^2 + \Delta^2)^{1/2}$ is sketched as a function of $\mathscr{E}$ in Fig. 6.1. For $\varepsilon \ll \Delta$ the curves are quadratic, while for $\varepsilon \gg \Delta$ they tend to the straight lines given by degenerate perturbation theory. The eigenfunctions come from (1). Substitution of the limiting values of λ shows that for the upper energy curve the eigenfunction tends to u_{210} as ε/Δ tends to zero, and to $(u_{210} - u_{200})/\sqrt{2}$ as ε/Δ tends to infinity. The corresponding limiting eigenfunctions for the lower energy curve are u_{200} and $(u_{210} + u_{200})/\sqrt{2}$.

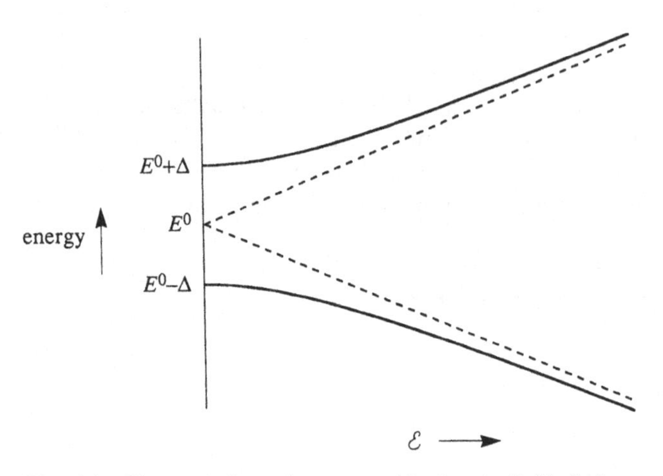

Fig. 6.1. *The variation of energy with electric field $\mathscr{E}$ for near degeneracy of the unperturbed states u_{210}, u_{200} of hydrogen. For $\varepsilon = 3a_0e\mathscr{E} \ll \Delta$, the curves are quadratic, while for $\varepsilon \gg \Delta$, they are linear. The dashed lines show the result for $\Delta = 0$.*

(d) A wavenumber separation $\tilde{\nu}$ corresponds to an energy separation

$$2\Delta = ch\tilde{\nu}. \tag{4}$$

For

$$\varepsilon = 3a_0e\mathscr{E} = \Delta, \tag{5}$$

$$\mathscr{E} = \frac{ch\tilde{\nu}}{6ea_0} = 1.4 \times 10^5 \text{ V m}^{-1}. \tag{6}$$

Comments

(1) This problem illustrates the procedure in a perturbation calculation when two or more states are almost degenerate. The first-order energies

are the eigenvalues of the matrix of $H^{(1)}$ for the quasi-degenerate states, and the eigenvectors of the matrix are the required linear combinations of the state functions that must be used for the unperturbed wave function. In this way the perturbed wave functions will tend smoothly to the unperturbed functions as the perturbation $H^{(1)}$ tends to zero.

(2) The present calculation for the splitting of the $n = 2$ states in hydrogen is not quite correct, due to spin–orbit effects. The Dirac theory of the electron, which incorporates the spin, shows that there are two energy levels for the $n = 2$ states, but the corresponding states are not as simple as in the present problem. The four $p_{3/2}$ states have the higher energy, while the two $p_{1/2}$ states and the two $s_{1/2}$ states have the lower energy. (Actually, there are further refinements. A quantum electrodynamical effect, known as the *Lamb shift*, causes the $s_{1/2}$ states to be raised by 3.5 m^{-1} above the $p_{1/2}$ states, and the spin of the nucleus causes all eight states to be further split by about 0.5 m^{-1}. But we shall ignore both these effects here.)

The difference between the two energy levels is 36 m^{-1} as stated in the problem. The method of the present problem can be used to calculate the first-order energies. The calculation is straightforward, though somewhat laborious. The matrix of $H^{(0)} + H^{(1)}$ is calculated on the basis of the eight states. Fortunately, the 8×8 matrix factorises into two identical 3×3 matrices, and two one-element matrices. The results for the energy eigenvalues are shown in Fig. 6.2. As can be seen, there is no linear effect for the upper unperturbed energy level, because it has only $l = 1$ states.

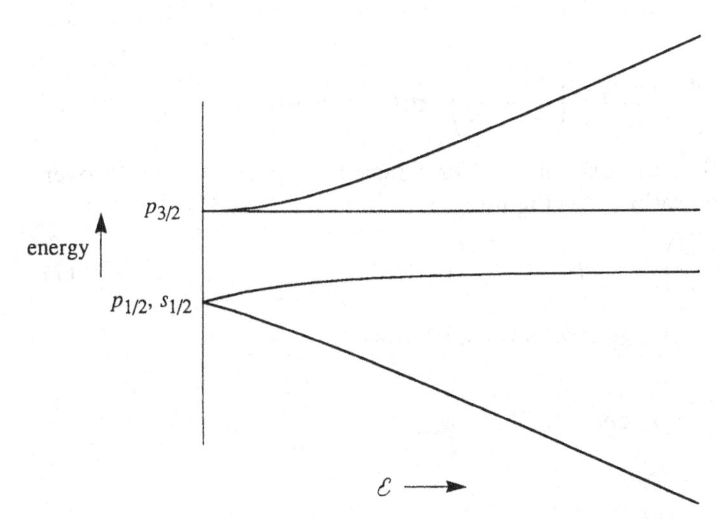

Fig. 6.2. *The Stark effect in hydrogen when spin–orbit forces are included. Each energy curve is doubly degenerate.*

However, the lower energy contains both $l = 1$ and $l = 0$ states, and there is therefore a linear effect. Each of the curves in the diagram is doubly degenerate, corresponding to a space state, which determines the energy, multiplied by the spin state α or β.

6.7 (a) For a point nucleus the potential is

$$V^{(0)} = -\frac{e^2}{4\pi\varepsilon_0}\frac{1}{r}. \tag{1}$$

For the thin uniform spherical shell the potential is

$$V = -\frac{e^2}{4\pi\varepsilon_0}\frac{1}{r} \quad r > b, \quad V = -\frac{e^2}{4\pi\varepsilon_0}\frac{1}{b} \quad r < b. \tag{2}$$

The perturbation potential is therefore

$$\begin{aligned} V^{(1)} = V - V^{(0)} &= \frac{e^2}{4\pi\varepsilon_0}\left(\frac{1}{r} - \frac{1}{b}\right) \qquad r < b, \\ &= 0 \qquad\qquad\qquad\qquad\quad r > b. \end{aligned} \tag{3}$$

The first-order correction to the energy of the ground state is

$$E^{(1)} = \int u_{1s}^* V^{(1)} u_{1s}\, \mathrm{d}\tau, \tag{4}$$

where

$$u_{1s} = (1/\pi a_0^3)^{1/2} \exp(-r/a_0). \quad \text{(See pp. 52 and 54.)} \tag{5}$$

Thus

$$E^{(1)} = \frac{4}{a_0^3}\frac{e^2}{4\pi\varepsilon_0}\int_0^b r^2\left(\frac{1}{r} - \frac{1}{b}\right)\exp(-2r/a_0)\,\mathrm{d}r. \tag{6}$$

Since $b/a_0 \approx 10^{-5}$, the exponential term may be replaced by unity over the range of integration, and the integral becomes

$$\int_0^b \left(r - \frac{r^2}{b}\right)\mathrm{d}r = \tfrac{1}{6}b^2. \tag{7}$$

The ground-state energy for a point nucleus is

$$E^{(0)} = -\frac{e^2}{4\pi\varepsilon_0}\frac{1}{2a_0}. \tag{8}$$

From (6), (7), and (8)

$$\frac{E^{(1)}}{E^{(0)}} = -\frac{4b^2}{3a_0^2}. \tag{9}$$

The value of b is about 10^{-15} m; the Bohr radius $a_0 = 0.53 \times 10^{-10}$ m. These values give $E^{(1)}/E^{(0)} \sim 5 \times 10^{-10}$. So the fractional change in energy is negligible in this case.

(b) The Bohr theory gives the result that for a particle of mass m in orbit round a nucleus of charge Ze, the radius is proportional to $1/Zm$. The Schrödinger equation gives the same result. Therefore, for the μ^- atom, the quantity a_0 must be replaced by

$$a = \frac{a_0}{Z}\frac{m_e}{m_\mu}, \tag{10}$$

where m_e and m_μ are the respective masses of the electron and the μ-meson. $Z = 82$ for lead; $m_\mu = 207m_e$. Therefore $a = a_0/(82 \times 207)$ for a μ-meson in lead. The radius of the nucleus varies roughly as $A^{1/3}$, where A is the mass number, which is about 208 for lead. Thus b is about 6 times larger for lead than for hydrogen. Inserting these factors in (9) gives $E^{(1)}/E^{(0)}$ of the order of unity for a μ-meson in lead. Since perturbation theory is valid only if $E^{(1)}/E^{(0)} \ll 1$, the calculated numerical value would not be valid in this case – it merely indicates that $E^{(1)}$ is of the same order as $E^{(0)}$. Experiments confirm this conclusion. For example, the measured value of the energy of the X-rays corresponding to the K_α transition for a μ-meson in lead is about 6 MeV, compared to the theoretical value for a point nucleus of 16 MeV.

6.8 (a) The upper limit of the energy of the ground state is the minimum value of the expectation value of the energy as β varies. Denote the trial wave function by

$$\psi = A \exp(-\beta r), \tag{1}$$

where A is a constant. Its value is obtained from the normalising condition

$$\int_{\substack{\text{all}\\\text{space}}} |\psi|^2 \, d\tau = 1. \tag{2}$$

Since ψ depends only on r, we put

$$d\tau = 4\pi r^2 \, dr, \tag{3}$$

and (2) becomes

$$4\pi A^2 \int_0^\infty r^2 \exp(-2\beta r) \, dr = \frac{8\pi A^2}{(2\beta)^3} = 1. \tag{4}$$

In evaluating the integral we have used (6.5.3). Thus

$$A^2 = \frac{\beta^3}{\pi}. \tag{5}$$

The Hamiltonian for the hydrogen atom is

$$H = -\frac{\hbar^2}{2m_e}\nabla^2 - \frac{e^2}{4\pi\varepsilon_0}\frac{1}{r}. \tag{6}$$

The expectation value of the energy for the function ψ is

$$\langle E \rangle = 4\pi A^2 \int_0^\infty \exp(-\beta r)\left(-\frac{\hbar^2}{2m_e}\nabla^2 - \frac{e^2}{4\pi\varepsilon_0}\frac{1}{r}\right)\exp(-\beta r) r^2\,\mathrm{d}r. \tag{7}$$

Since $\exp(-\beta r)$ depends only on r, we need only the r-dependent part of the operator ∇^2, which is $\mathrm{d}^2/\mathrm{d}r^2 + (2/r)\,\mathrm{d}/\mathrm{d}r$. Thus

$$\nabla^2 \exp(-\beta r) = \left(\beta^2 - \frac{2\beta}{r}\right)\exp(-\beta r). \tag{8}$$

Inserting (5) and (8) in (7) gives

$$\begin{aligned}\langle E \rangle &= 4\beta^3 \left\{-\frac{\hbar^2}{2m_e}\beta^2 \int_0^\infty r^2 \exp(-2\beta r)\,\mathrm{d}r \right.\\ &\qquad \left. + \left(\frac{\beta\hbar^2}{m_e} - \frac{e^2}{4\pi\varepsilon_0}\right)\int_0^\infty r \exp(-2\beta r)\,\mathrm{d}r\right\}\\ &= -\frac{\hbar^2}{2m_e}\beta^2 + \frac{\hbar^2}{m_e}\beta^2 - \frac{e^2}{4\pi\varepsilon_0}\beta\\ &= \frac{\hbar^2}{2m_e}\beta^2 - \frac{e^2}{4\pi\varepsilon_0}\beta. \end{aligned} \tag{9}$$

Then

$$\begin{aligned}\frac{\mathrm{d}\langle E \rangle}{\mathrm{d}\beta} &= \frac{\hbar^2\beta}{m_e} - \frac{e^2}{4\pi\varepsilon_0}\\ &= 0, \quad \text{when} \quad \beta = \frac{m_e e^2}{4\pi\varepsilon_0\hbar^2}. \end{aligned} \tag{10}$$

Inserting this value in (9) gives

$$\langle E \rangle_{\min} = -\frac{m_e}{2\hbar^2}\left(\frac{e^2}{4\pi\varepsilon_0}\right)^2 \tag{11}$$

$$= -\frac{e^2}{8\pi\varepsilon_0}\frac{1}{a_0}, \tag{12}$$

where

$$a_0 = \frac{4\pi\varepsilon_0}{e^2}\frac{\hbar^2}{m_e} \quad \text{is the Bohr radius.} \tag{13}$$

(b) The expression in (11) is in fact the correct value of the ground-state energy of the hydrogen atom. The reason for this is that the trial function $\exp(-\beta r)$ is the correct form of the ground-state wave function. As can be seen, the expectation value of the energy is a minimum when $\beta = 1/a_0$, and $\exp(-r/a_0)$ is the actual wave function for the ground state of the hydrogen atom.

6.9 Denote the attractive potential of arbitrary shape by V (a function of x) and the attractive square-well potential by V_s. Then the Hamiltonians for the two potentials are

$$H = T + V, \quad \text{and} \quad H_s = T + V_s. \tag{1}$$

where T is the kinetic energy operator.

Let ψ_s be an eigenfunction of H_s with energy E_s, which is negative. From the result of Problem 3.3 there must be at least one such eigenfunction. Now

$$\int \psi_s^*(T + V_s)\psi_s \, dx = E_s. \tag{2}$$

Let E_0 be the ground-state energy for the potential V. By the variational principle, the expectation value of the energy for any wave function is greater than or equal to E_0. So

$$\int \psi_s^*(T + V)\psi_s \, dx \geqslant E_0. \tag{3}$$

From (2) and (3)

$$\int \psi_s^*(V - V_s)\psi_s \, dx \geqslant E_0 - E_s. \tag{4}$$

However shallow the square well there is always one bound state for it. Therefore, since V is a negative function of x, we can make $V - V_s$ negative for all x. In which case $\int |\psi_s|^2(V - V_s)\, dx$ is negative. So, from (4)

$$E_0 - E_s \leqslant \text{negative quantity.} \tag{5}$$

Since E_s is negative, it follows from (5) that E_0 is negative, i.e. there is at least one bound state for the potential V.

7
Identical particles, multielectron atoms

Summary of theory

1 Symmetric and antisymmetric states

Suppose we have a system of two particles (1) and (2), with (1) in a normalised single-particle state u_a, and (2) in a normalised single-particle state u_b. The state of the system is

$$\psi(1, 2) = u_a(1)u_b(2). \tag{1}$$

If the particles are exchanged, the state of the system becomes

$$\psi(2, 1) = u_a(2)u_b(1), \tag{2}$$

which, in general, may be quite different from the original state function. However, if the particles are identical bosons (particles with zero or integral spin), the state function must be symmetric, i.e. it must remain the same when the two particles are interchanged. If they are identical fermions (particles with half-integral spin), the state function must be antisymmetric, i.e. it must change sign when the two particles are interchanged. If there are more than two particles, the state function of the system must be symmetric for bosons (or antisymmetric for fermions) for the interchange of each pair of particles in the system.

We can readily construct symmetric and antisymmetric state functions from a function such as (1). Thus

$$\Psi(1, 2)_S = \{\psi(1, 2) + \psi(2, 1)\}/\sqrt{2}$$

is a symmetric function, (3)

$$\Psi(1, 2)_A = \{\psi(1, 2) - \psi(2, 1)\}/\sqrt{2}$$

is an antisymmetric function. (4)

The $1/\sqrt{2}$ factor is to preserve normalisation. Note that, if the two particles are in the same single-particle state, i.e. $a = b$ in (1), we can only have a symmetric state, because in this case

$$\psi(1, 2) = \psi(2, 1), \quad \text{and} \quad \Psi(1,2)_A = 0, \tag{5}$$

which corresponds to no state.

For three identical particles we use a similar notation. The state of the system when particle (1) is in a state u_a, particle (2) is in a state u_b, and particle (3) is in a state u_c, is written as

$$\psi(1, 2, 3) = u_a(1)u_b(2)u_c(3). \tag{6}$$

Similarly

$$\psi(2, 3, 1) = u_a(2)u_b(3)u_c(1), \tag{7}$$

and so on. In other words, we adopt the convention that we keep the states in the same order a, b, c, and the order of the numbers in the argument of ψ tells us which particle is in each of the three states.

For three particles, the symmetric and antisymmetric functions are

$$\Psi(1, 2, 3)_S = [\{\psi(1, 2, 3) + \psi(2, 3, 1) + \psi(3, 1, 2)\} \\ + \{\psi(2, 1, 3) + \psi(1, 3, 2) + \psi(3, 2, 1)\}]/\surd(3!). \tag{8}$$

$$\Psi(1, 2, 3)_A = [\{\psi(1, 2, 3) + \psi(2, 3, 1) + \psi(3, 1, 2)\} \\ - \{\psi(2, 1, 3) + \psi(1, 3, 2) + \psi(3, 2, 1)\}]/\surd(3!). \tag{9}$$

For the symmetric function, all six terms have the positive sign. For the antisymmetric function, the rule for finding the sign of a particular term is to change pairs of numbers until the sequence 1, 2, 3 is obtained. The sign is positive or negative depending on whether the number of changes is even or odd. For example,

$$2, 3, 1 \rightarrow 3, 2, 1 \rightarrow 1, 2, 3 \quad \text{2 changes} \quad \psi(2, 3, 1) \text{ is positive,}$$

$$2, 1, 3 \rightarrow 1, 2, 3 \quad \text{1 change} \quad \psi(2, 1, 3) \text{ is negative.}$$

The procedure is readily generalised to n particles. We write down the $n!$ permutations of the order of the numbers in $\psi(1, 2, \ldots n)$, and add all the terms for the symmetric function. For the antisymmetric function, we add or subtract the terms according to whether the number of pair interchanges necessary to reach the sequence $1, 2, \ldots n$ is even or odd.

2 Symmetry of angular momentum states for two identical particles

The results in the first part of Table 5.1 (p. 96) show that, for two identical particles with spin $s_1 = s_2 = \frac{1}{2}$, the resultant states $S = 1$ are symmetric, and the state $S = 0$ is antisymmetric with respect to the interchange of the two particles. We now show that this is part of a general pattern. Consider the case of two identical particles with angular momentum $l_1 = l_2 = 1$, giving a resultant $L = 2, 1, 0$. It is easy to see that the $L = 2$ functions are symmetric. Clearly, in the nomenclature of Problem 5.7, the function with $L = 2$, $M_L = 2$ is

$$\Phi_{2,2} = \phi_1\phi_1, \tag{10}$$

which is symmetric. Moreover, the function Φ_{21} may be obtained by applying the lowering operator $L_- = L_{1-} + L_{2-}$ to both sides of (10). Now the lowering operator is itself symmetric with respect to the interchange of the two particles, so when it operates on a function it leaves its symmetry unchanged. Therefore, since $\Phi_{2,2}$ is symmetric, Φ_{21} is also symmetric. In the same way, $\Phi_{2,0}$ has the same symmetry as Φ_{21}, and so on. So, for a given value of L, the symmetry of the Φs is the same for all the values of M_L. The symmetry depends only on the value of L.

The function Φ_{21} is given by

$$\Phi_{21} = (\phi_0\phi_1 + \phi_1\phi_0)/\sqrt{2}. \tag{11}$$

The function Φ_{11}, being orthogonal to Φ_{21}, is

$$\Phi_{11} = (\phi_0\phi_1 - \phi_1\phi_0)/\sqrt{2}, \tag{12}$$

which is an antisymmetric function. So by the previous argument the three $L = 1$ functions are antisymmetric. The reasoning may be continued. The function Φ_{00}, which is orthogonal to both Φ_{20} and Φ_{10}, is symmetric.

The general result is as follows. If we have two identical particles with angular momentum quantum numbers $j_1 = j_2 = j$, then the resultant quantum number J takes values $J = 2j, 2j - 1, 2j - 2, \ldots 0$. The states $J = 2j$ are symmetric, the states $J = 2j - 1$ are antisymmetric, and the symmetry continues to alternate as J decreases. Note that the present results apply whatever kind of angular momentum the quantum number j represents, whether it be orbital, spin, or a combination of the two.

3 Multielectron atoms

By multielectron atoms in the present context we mean atoms with two or more electrons in the unfilled or valency shell. The calculation of the energy levels of these atoms is complex. The forces involved are the electrostatic interactions between the nucleus and the electrons, and between the electrons themselves. In addition there are the magnetic spin–orbit forces. The calculation proceeds in two stages.

In the first, known as the *central field approximation*, all the magnetic interactions are neglected, and the actual electrostatic interactions are replaced by a spherically symmetrical potential, which, for a given electron, represents the electrostatic force due to the nucleus and the average of those due to all the other electrons. The eigenfunctions of this zero-order Hamiltonian $H^{(0)}$ are specified by the *electronic configuration*, which gives the number of electrons in each shell. For example, the electronic configuration of oxygen in the ground state is $(1s)^2(2s)^2(2p)^4$, meaning that there are two electrons in the $1s$ shell ($n = 1$, $l = 0$), two in the $2s$ shell ($n = 2$, $l = 0$), and four in the $2p$ shell ($n = 2$, $l = 1$). All the electrons in the same shell have the same energy in the central field approximation.

The second stage of the calculation uses degenerate perturbation theory, the Hamiltonian $H^{(1)}$ being the sum of the corrections V' to the electrostatic terms in $H^{(0)}$, and the magnetic interactions P. The form of the resulting eigenfunctions follows from the commutation relations, and depends on the relative magnitudes of V' and P.

(a) LS coupling

If $P \ll V'$, we have what is known as *LS coupling*. The operators L^2, L_z, S^2, S_z, which commute with V', also commute with $H^{(1)}$. (This statement and the ones that follow are strictly true only for the extreme case of $P = 0$.) Thus the state functions (eigenfunctions of the Hamiltonian) are characterised by single values of the quantum numbers L, M_L, representing the total orbital angular momentum of the electrons, and its z component. We say that L and M_L are *good quantum numbers*. Similarly, S and M_S are good quantum numbers. On the other hand, the operators J_1^2, and J_2^2, representing the total (orbital plus spin) angular momentum of each of the valency electrons, do not commute with $H^{(1)}$. (For simplicity we restrict the discussion to the case of two electrons in the valency shell.) The state functions are, in general, mixtures of states with different values of j_1, j_2. The condition for LS coupling, i.e. $P \ll V'$, applies to the majority of atoms and to all light ones. A set of states with the same L, S values and different values of J is known as an *LS multiplet*. If P were zero, all the states in the multiplet would have the same energy, but the small non-zero P causes a small splitting of the energies.

The values of L, S, J for the ground state are given by *Hund's rules*. These are, first maximise S, then maximise L. Thirdly, $J = |L - S|$ if the shell is less than half full, and $J = L + S$ if the shell is more than half full. The maximum values of S and L are subject to the condition that no two electrons may have the same pair of values for m_s and m_l. If the shell is exactly half full, the rules lead to the result $L = 0$, $J = S$.

The notation used to specify the values of the quantum numbers L, S, J of a state in LS coupling is ${}^{2S+1}L_J$, where L is the conventional letter for the L value, i.e. S for $L = 0$, P for $L = 1$, D for $L = 2$, and so on. Thus ${}^2P_{3/2}$ represents the values $L = 1$, $S = \frac{1}{2}$, $J = \frac{3}{2}$.

(b) *jj* coupling

When $P \gg V'$, we have *jj coupling*. The operators J_1^2, J_{1z}, J_2^2, J_{2z} commute with $H^{(1)}$ (strictly only when $V' = 0$), while the operators L^2, L_z, S^2, S_z do not. Thus the state functions are characterised by single values of the quantum numbers j_1, j_2, but are mixtures of states with different values of L, M_L, S, M_S. A set of states with the same j_1, j_2 values and different values of J is known as a *jj multiplet*. If V' were zero, all the states in the multiplet would have the same energy, but the small non-zero V' causes a small splitting of the energies. Only a few atoms, with high values of the atomic number, obey *jj* coupling. The quantum numbers of a state in *jj* coupling are specified as (j_1, j_2, J).

(c) Intermediate coupling

For a few atoms $P \sim V'$, a situation known as *intermediate coupling*. Here, none of the operators L, M_L, S, M_S, J_1^2, J_{1z}, J_2^2, J_{2z} commutes with the Hamiltonian. The operator J^2, representing the resultant overall angular momentum of the electrons, commutes with $H^{(1)}$, whatever the relative values of V' and P.

Problems

7.1 (a) Show that for a system of two identical particles, each of which can be in one of n quantum states, there are

$$\tfrac{1}{2}n(n+1) \quad \text{symmetric,}$$

and

$$\tfrac{1}{2}n(n-1) \quad \text{antisymmetric}$$

states of the system.

(b) Show that, if the particles have spin I, the ratio of symmetric to antisymmetric spin states is $(I+1){:}I$.

7.2 Two non-interacting particles, with the same mass m, are in a one-dimensional potential which is zero along a length $2a$, and infinite elsewhere.

(a) What are the values of the four lowest energies of the system?.

(b) What are the degeneracies of these energies if the two particles are

(i) identical, with spin $\frac{1}{2}$,

(ii) not identical, but both have spin $\frac{1}{2}$,

(iii) identical, with spin 1.

7.3 Two identical non-interacting particles are in an isotropic harmonic potential. Show that the degeneracies of the three lowest energy levels are

(a) 1, 12, 39, if the particles have spin $\frac{1}{2}$,

(b) 6, 27, 99, if the particles have spin 1.

7.4 The ρ^0 meson has spin 1, and the π^0 meson has spin 0. Show that the decay

$$\rho^0 \to \pi^0 + \pi^0$$

is impossible.

7.5 The π^- meson interacts with the deuteron from an s orbital state to form two neutrons.

(a) Show that the neutrons are in the state $L = 1$, $S = 1$, $J = 1$.

(b) Hence deduce the intrinsic parity of the π^-.

[The π^- has spin 0, and the deuteron has spin 1.]

7.6 Calculate the degeneracy, and list the possible ${}^{2S+1}L_J$ values for each of the following electronic configurations.

(a) $2s2p$, (b) $2p3p$, (c) $(2p)^2$, (d) $(3d)^{10}$, (e) $(3d)^9$.

For each configuration, verify that the sum of the number of states of each $^{2S+1}L_J$ combination is equal to the degeneracy of the configuration.

7.7 The electronic configuration of the ground state of nitrogen is $(1s)^2(2s)^2(2p)^3$.

(a) What is the degeneracy of this electronic configuration?

(b) List the permitted products of single-electron states for the electrons in the $2p$ shell in order of decreasing M_L value.

(c) What are the possible $^{2S+1}L_J$ values for the electronic configuration?

(d) Which set has the lowest energy?

7.8 The lead atom obeys jj coupling, and has two electrons in the $6p$ shell in the ground state. What are the possible j_1, j_2, J values for this electronic configuration?

7.9 In the shell model, the ground state of the nucleus ^{19}O is represented by three $d_{5/2}$ neutrons outside closed shells. Given that the nucleons obey jj coupling, calculate the possible values of the spin of the ground state of the nucleus.

7.10 In the ortho form of the hydrogen molecule the spin wave function of the two nuclei is symmetric, and in the para form it is antisymmetric. The internal motion of the molecule may be represented by that of a rigid rotator. The eigenstates of the latter are the spherical harmonics Y_{lm}, and the corresponding energy eigenvalues are $l(l+1)\hbar^2/2I$, where I is the moment of inertia of the molecule about an axis through its centre of mass, perpendicular to the line of the nuclei.

(a) Show that, in an equilibrium mixture of light hydrogen at temperature T, the ratio of n_p, the number of para molecules, to n_o, the number of ortho molecules, is

$$\frac{n_p}{n_o} = \frac{1}{3}\frac{\sum_{l\ \mathrm{even}}(2l+1)\exp\{-l(l+1)x\}}{\sum_{l\ \mathrm{odd}}(2l+1)\exp\{-l(l+1)x\}},$$

where

$$x = \frac{\hbar^2}{2I}\frac{1}{k_B T}; \quad k_B \text{ is the Boltzmann constant.}$$

(b) At a temperature of 20.0 K, an equilibrium mixture of light hydrogen consists of 99.83% of the para form. Calculate the separation distance of the two nuclei in the molecule.

(c) Assuming the distance between the two nuclei is the same for light and heavy hydrogen, calculate the percentage of ortho molecules in an equilibrium mixture of heavy hydrogen at the same temperature.

7.11 What are the degeneracies of the two lowest rotational energy levels of HD (the hydrogen–deuterium molecule)?

7.12 The trebly ionised ion of praseodymium has two electrons in the $4f$ shell.

(a) Use Hund's rules to calculate the values of L, S, J for the ground state of the ion, and find the value of the Landé g-factor.

(b) Show that, when a magnetic field $B = 1\,\mathrm{T}$ is applied to a praseodymium salt at temperature $T = 300\,\mathrm{K}$, the thermal average of the component of the magnetic moment of a praseodymium ion in the direction of **B** is, to good approximation,

$$\bar{\mu} = \tfrac{1}{3}g^2\mu_\mathrm{B}^2\frac{J(J+1)B}{k_\mathrm{B}T},$$

where μ_B is the Bohr magneton.

(c) Estimate the magnetic susceptibility of the salt per mole of praseodymium for these values of B and T.
[Hint: Use the result that the probability that an ion has a quantum number M_J is proportional to the Boltzmann factor $\exp(-E/k_\mathrm{B}T)$, where E is the magnetic energy for the value M_J. Assume that, at the temperature of the problem, $k_\mathrm{B}T$ is small compared to the LS multiplet spacing.]

Solutions

7.1 (a) If there are n single-particle states, there are n symmetric states in which both particles are in the same state. The number of symmetric states in which the particles are in different states is equal to the number of ways we can select two objects from n, i.e. $\frac{1}{2}n(n-1)$. So the total number of symmetric states is

$$n + \tfrac{1}{2}n(n-1) = \tfrac{1}{2}n(n+1). \tag{1}$$

The number of antisymmetric states is $\frac{1}{2}n(n-1)$, because states in which both particles are in the same state are excluded.

(b) If the particles have spin I, there are $n = 2I + 1$ single-particle states (corresponding to the $2I + 1$ different m values that give the component of spin angular momentum in any direction in space). Therefore the number of symmetric states is

$$n_s = (I+1)(2I+1), \tag{2}$$

and the number of antisymmetric states is

$$n_a = I(2I+1). \tag{3}$$

The ratio of symmetric to antisymmetric states is thus $(I+1)/I$.

7.2 (a) The single-particle space wave functions must be zero at $x = 0$ and $x = 2a$. They have the form

$$\psi(x) = \frac{1}{\sqrt{a}} \sin(n\pi x/2a), \tag{1}$$

where n is a positive integer. The corresponding energy is $n^2 E_0$, where

$$E_0 = \frac{\pi^2 \hbar^2}{8ma^2} \tag{2}$$

– see Problem 3.1. Denote a state for the two particles by n_1, n_2, which stands for

$$\psi(x_1, x_2) = \frac{1}{a} \sin(n_1 \pi x_1/2a) \sin(n_2 \pi x_2/2a). \tag{3}$$

The state has energy

$$E = (n_1^2 + n_2^2) E_0. \tag{4}$$

The four lowest energies are thus

n_1, n_2	E/E_0
1, 1	2
2, 1	5
2, 2	8
3, 1	10

(b) (i) The state function may be expressed as a product of a space function and a spin function. Since the particles are identical fermions, the state function must be antisymmetric. Thus if the space function is symmetric the spin function must be antisymmetric, and vice versa. For $I = \frac{1}{2}$, the results of the last problem show that there are three symmetric and one antisymmetric spin functions. (They are the functions listed in the first half of Table 5.1. As can be seen from the table, the $S = 1$ functions are symmetric, and the $S = 0$ function is antisymmetric.)

The space state 1,1 must be symmetric. Therefore the overall state is the 1,1 space state multiplied by the antisymmetric $S = 0$ state. The degeneracy is thus one. The specific form of the normalised state function is

$$\psi = \frac{1}{\sqrt{2a}} \sin(\pi x_1/2a) \sin(\pi x_2/2a)(\alpha\beta - \beta\alpha), \tag{5}$$

where the first α or β in a product gives the spin state of the first particle, and the second α or β gives that of the second particle.

The space state 2,1 can give either a symmetric or an antisymmetric function. The former must be paired with the $S = 0$ spin state; the latter with one of the three $S = 1$ spin states. So the degeneracy of the second energy value is four. As examples, the specific forms of two of the four state functions are

$$\psi = \frac{1}{2a}\{\sin(2\pi x_1/2a) \sin(\pi x_2/2a) + \sin(\pi x_1/2a) \sin(2\pi x_2/2a)\}(\alpha\beta - \beta\alpha), \tag{6}$$

$$\psi = \frac{1}{\sqrt{2a}}\{\sin(2\pi x_1/2a) \sin(\pi x_2/2a) - \sin(\pi x_1/2a) \sin(2\pi x_2/2a)\}\alpha\alpha. \tag{7}$$

In general, the degeneracy of an energy corresponding to a space state n_1, n_2 is one if $n_1 = n_2$, and four if $n_1 \neq n_2$.

(ii) If the particles are not identical, there is no restriction on the symmetry of either the space or spin part of the state function. If $n_1 \neq n_2$, there are two space states with the same energy. Thus

$$\frac{1}{a}\sin(2\pi x_1/2a)\sin(\pi x_2/2a) \quad \text{and} \quad \frac{1}{a}\sin(\pi x_1/2a)\sin(2\pi x_2/2a)$$

are permitted space states with the same energy. Each may be coupled with any one of the spin functions $\alpha\alpha$, $\alpha\beta$, $\beta\alpha$, $\beta\beta$, giving an overall degeneracy of eight. If $n_1 = n_2$, there is only one space state, so the degeneracy is four.

(iii) The state function of a system of identical particles with integral spin is symmetric. So a symmetric space function must have a symmetric spin function, and vice versa. From (7.1.2) and (7.1.3), for two identical particles with $I = 1$, there are six symmetric and three antisymmetric spin functions. The space function 1, 1, being symmetric, must be paired with one of the six symmetric spin functions. As before, the space state 2, 1 can give a symmetric or an antisymmetric function, so the spin function can be symmetric or antisymmetric, giving a degeneracy of nine. Similarly the states 2, 2 and 3, 1 have respective degeneracies six and nine.

The results are summarised in the table.

Energy E/E_0	Degeneracy (i)	(ii)	(iii)
2	1	4	6
5	4	8	9
8	1	4	6
10	4	8	9

7.3 As shown in Problem 4.10, the state of a single particle in an isotropic harmonic potential can be characterised by a trio of non-negative integers n_1, n_2, n_3. The state has energy $(n + \frac{3}{2})\hbar\omega$, where $n = n_1 + n_2 + n_3$. The three lowest energies correspond to

n	n_1	n_2	n_3
0	0	0	0
1	1	0	0
2	2	0	0
	1	1	0

Each trio of integers (apart from the first) may be rearranged to give another state with the same energy. For two non-interacting particles, the space state is a product of two single-particle states. The energy of the state is the sum of the two single-particle energies.

(a) For spin $\frac{1}{2}$ particles, the overall wave function is antisymmetric. The ground state corresponds to both particles in the state 0,0,0, which we write as $(0, 0, 0)(0, 0, 0)$. Since this is a symmetric function, the spin state is the singlet antisymmetric state. So the ground-state energy is non-degenerate.

The next energy level corresponds to one particle in the state $1, 0, 0$ and one in the state $0, 0, 0$, i.e. to $(1, 0, 0)(0, 0, 0)$. Since the two particles are in different space states, we can make the space state for the pair either symmetric or antisymmetric. As we saw in part b(i) of the last problem, this gives rise to four space–spin functions. But the integer 1 in the space function can also occur in the second or third position. So the total degeneracy of this energy is $3 \times 4 = 12$.

The next energy level can be obtained with the following combinations of single-particle states.

$$
\begin{aligned}
&(2, 0, 0)(0, 0, 0), \\
&(1, 1, 0)(0, 0, 0), \\
&(0, 1, 0)(0, 0, 1) \quad \text{(the two 1s in different positions)}, \\
&(1, 0, 0)(1, 0, 0) \quad \text{(the two 1s in the same position)}. \qquad (1)
\end{aligned}
$$

The first three combinations correspond to the two particles being in different space states. Each therefore gives rise to 12 states as before. The fourth combination corresponds to both particles in the same space state, and therefore to a single space-spin state. But we multiply by 3, because the two 1s can be in each of the three positions. So the total degeneracy of this energy level is $(3 \times 12) + 3 = 39$.

(b) For two identical particles with spin 1, there are six symmetric and three anti-symmetric spin functions. The overall space–spin function must be symmetric. For the ground state there is a single symmetric space function, which is combined with one of the six symmetric spin functions, giving a degeneracy of six.

For the next energy level, the state $(1, 0, 0)(0, 0, 0)$ can give a symmetric or an antisymmetric space function, which is combined with one of the nine spin functions. Multiplying by 3 for the three positions of the 1 gives an overall degeneracy of 27. For the next energy level, each of the first three combinations listed in (1) gives 27 space–spin states. For the last combination in (1) there are six space–spin states, just as for the ground-state energy, which is multiplied by 3 for the three positions of the 1. So the overall degeneracy of the energy level is $(3 \times 27) + (3 \times 6) = 99$.

7.4 The ρ^0 meson has spin 1, and the π^0 meson has spin 0. Therefore, if the decay were to occur, the two π^0 mesons would have to be in an $L = 1$ orbital state in order to conserve angular momentum. However, an $L = 1$ state has odd parity, which is an antisymmetric space function. (For a system of two particles, the parity and symmetry of the space function are the same, because reflecting the wave function through the origin interchanges the two particles. Reflecting a spin through the origin leaves it unchanged, so the parity of a spin function is always positive.) Since the mesons have zero spin, the spin state function is symmetric. Therefore the overall state function is antisymmetric. But the π^0 meson is a boson and must have a symmetric state function. The decay is therefore impossible.

7.5 (a) The reaction is

$$\pi^- + \mathrm{d} \rightarrow \mathrm{n} + \mathrm{n}. \tag{1}$$

The π^- has spin 0, and the deuteron has spin 1. So the total spin of the left-hand side is $S = 1$. The particles interact from an s-orbital state, i.e. $L = 0$. Therefore $J = 1$ for the left-hand side, and, by conservation of angular momentum, this is the J value for the right-hand side.

The neutron has spin $\frac{1}{2}$, so the spin of the right-hand side is $S = 1$ or 0. We can rule out $S = 0$, because that is an antisymmetric spin state, which requires a symmetric space state, i.e. an even value of L. This would give an even value of J, so angular momentum would not be conserved. The symmetric spin state $S = 1$ requires an antisymmetric space state, i.e. an odd value of L. The only odd value, which combined with $S = 1$ gives $J = 1$, is $L = 1$.

(b) The deuteron consists of a proton and a neutron in an $L = 0$ state, which has positive parity. Therefore the parity of the deuteron is positive. (The intrinsic parities of the proton and the neutron are positive.) Since the π^- and the deuteron interact from an $L = 0$ state, the overall parity of the left-hand side is the same as the intrinsic parity of the π^-. The right-hand side consists of two neutrons in an $L = 1$ state; its parity is therefore negative. The reaction occurs via the strong nuclear force, which conserves parity. So the intrinsic parity of the π^- is negative.

7.6 (a) A $2s$ electron has $n = 2$, $l = 0$, $s = \frac{1}{2}$, with two states corresponding to $m_s = \pm\frac{1}{2}$. A $2p$ electron has $n = 2$, $l = 1$, $s = \frac{1}{2}$, with six states given by $m_l = 1, 0, -1$, $m_s = \pm\frac{1}{2}$. (The n value is not relevant to the number of single-particle states.) Each $2s$ state may be combined with a $2p$ state and an antisymmetric function formed from the combination. So the degeneracy of the electronic configuration is $2 \times 6 = 12$.

To obtain the L, S, J values, we first add the orbital angular momenta and the spin angular momenta separately. We have

$$l_1 = 0, l_2 = 1, s_1 = s_2 = \tfrac{1}{2}. \tag{1}$$

Therefore

$$L = 1, \quad S = 1 \text{ or } 0. \tag{2}$$

For $L = 1$, there are three states, corresponding to $M_L = 1, 0, -1$. Each of these corresponds to a pair of different single-particle states, with one electron having $m_l = 0$, and the other $m_l = 1, 0, -1$. For each pair of single-particle states we can construct a symmetric or an antisymmetric space state. The former is combined with the antisymmetric spin state $S = 0$, and the latter with one of the three symmetric $S = 1$ spin states. We therefore have

$$L = 1, S = 1, \quad \text{and} \quad L = 1, S = 0. \tag{3}$$

Once we have an antisymmetric L, S combination, there is no symmetry restriction on the values of J, which are given by (5.24). We thus have

$$L = 1, S = 1, J = 2, 1, 0, \tag{4}$$

$$L = 1, S = 0, J = 1, \tag{5}$$

or in standard nomenclature

$$^3P_2, {}^3P_1, {}^3P_0, {}^1P_1. \tag{6}$$

Each J value gives rise to $2J + 1$ states, corresponding to $M_J = J$, $J - 1$, ... $-J$. So the total number of states from (6) is $5 + 3 + 1 + 3 = 12$, which is the same as the degeneracy of the electronic configuration.

Comment

The sum of the $2J + 1$ values for each J is always equal the degeneracy of the electronic configuration. This is because each state L, S, J, M_J is a linear combination of the single-particle states m_{l_1}, m_{l_2}, m_{s_1}, m_{s_2}, and conversely. There must therefore be the same number of the two types of state. Note that in summing over the single-particle states to form a specific L, S, J, M_J state, only terms with

$$m_{l_1} + m_{l_2} + m_{s_1} + m_{s_2} = M_J \tag{7}$$

contribute. The situation is an extension of the addition of two angular momenta (Problem 5.7). The result that there is the same number of the two types of state is a useful check that all the J values (and no more) have been obtained.

(b) For the electronic configuration $2p3p$, each electron can be in one of six states. Each of the six states from $2p$ may be combined with one of the six from $3p$ to give an antisymmetric function. The degeneracy of the configuration is therefore $6 \times 6 = 36$. Note that some of the combinations correspond to both electrons having the same values for m_l and m_s. The electrons are not however in the same state, because they have different n values.

For two angular momenta $l_1 = l_2 = 1$, the resultant quantum number L can have values $L = 2, 1, 0$. We again use the result that for two electrons in different shells any combination of single-particle space states may be made into a symmetric or an antisymmetric function, and so any L may be combined with any S value. The possible L, S combinations and the resulting J values are listed in Table 7.1.

The states are

$$^3D_3,\ ^3D_2,\ ^3D_1,\ ^3P_2,\ ^3P_1,\ ^3P_0,\ ^3S_1,\ ^1D_2,\ ^1P_1,\ ^1S_0. \tag{8}$$

Table 7.1. *L, S, J values and number of states for the electronic configuration 2p3p.*

L	S	J	$2J+1$	sum of $2J+1$
2	1	3, 2, 1	7, 5, 3	15
1	1	2, 1, 0	5, 3, 1	9
0	1	1	3	3
2	0	2	5	5
1	0	1	3	3
0	0	0	1	1
			total	36

(c) There are six states for a p electron. The two electrons must be in different states, so there are 6×5 possible product states, but these are used in pairs to form antisymmetric functions. The degeneracy of an electronic configuration with two electrons in the same p shell is therefore $\frac{1}{2} \times 6 \times 5 = 15$.

As in (b), the possible L values are 2, 1, 0. However, because the electrons are now in the same shell, we cannot construct space functions of arbitrary symmetry for a given L value. The symmetry is governed by the value of L, just as the symmetry of the spin function is governed by the value of S. The symmetries are (p. 129)

$$\begin{array}{llll} & s\ \ a\ \ s & & s\ \ a \\ L = & 2\ \ 1\ \ 0, & S = & 1\ \ 0. \end{array} \tag{9}$$

To make the overall state function antisymmetric, we combine a symmetric space function with an antisymmetric spin function, and vice-versa. The L, S values and the resulting J values are given in Table 7.2.

The states are

$$^1D_2,\ ^3P_2,\ ^3P_1,\ ^3P_0,\ ^1S_0. \tag{10}$$

Table 7.2. *L, S, J values and number of states for the electronic configuration $(np)^2$.*

L	S	J	$2J+1$	sum of $2J+1$
2	0	2	5	5
1	1	2, 1, 0	5, 3, 1	9
0	0	0	1	1
			total	15

(d) A d shell corresponds to $l = 2$, with five values of m_l. Multiplying this by 2 for the spin states gives a total of ten. Thus the configuration $(3d)^{10}$ represents a full shell. It is non-degenerate, and the state is $^1S_0$. This is a general result for a full shell. It follows because each of the electrons must have a different pair of m_l, m_s values. In a full shell all the m_l values from l to $-l$ occur twice (once for each value of m_s), so their sum is zero. Similarly, the sum of the m_s values is zero. Thus the resultant quantum numbers M_L and M_S are zero. Therefore

$$L = S = 0. \tag{11}$$

Any other value for L or S would require a non-zero value for the quantum number M_L or M_S. The zero values for L and S give zero for J.

Comment

The fact that $L = S = J = 0$ for a full shell is a highly convenient result. It means that to find the L, S, J values for a given electronic configuration of an atom, we can ignore all the filled shells, which may contain a large number of electrons. We need consider only the relatively few electrons in the unfilled or valence shell.

(e) This configuration represents a set of electrons one short of a full shell. Since a full shell has zero angular momentum (both orbital and spin), it follows that, if one electron is removed from a full shell, the components of orbital and spin angular momentum of the remainder are minus those of the one that was removed. So the L, S, J values of the remainder are the same as if there were only one electron in the shell. Thus the configuration $(3d)^9$ has degeneracy 10, and its L, S, J values are

$$L = 2,\ S = \tfrac{1}{2},\ \text{with } J = \tfrac{5}{2}, \tfrac{3}{2}, \tag{12}$$

i.e. the states are $^2D_{5/2}$, $^2D_{3/2}$.

7.7 (a) The 1*s* and 2*s* shells are full and may therefore be ignored. There are six single-particle states for an electron in the 2*p* shell. Each of the three electrons in this shell must be in a different state. So the degeneracy of the electronic configuration is equal to the number of ways of choosing three difference states from six, i.e. it is

$$\frac{6!}{3!3!} = 20. \tag{1}$$

(b) The 20 product states are listed in the left-hand column of Table 7.3. The numbers in the products are the m_l values ($\bar{1}$ stands for -1), and α and β give the m_s values. Thus the first product function $1\alpha\, 0\alpha\, 1\beta$ represents the three single-particle states

$$m_l = 1,\ m_s = \tfrac{1}{2}, \quad m_l = 0,\ m_s = \tfrac{1}{2}, \quad m_l = 1,\ m_s = -\tfrac{1}{2}. \tag{2}$$

Table 7.3. *Product states and M_L, M_S values for the electronic configuration $(np)^3$.*

A tick indicates that the L, S, M_L, M_S state is a linear combination of the product states on the same line.

product of single-particle states	M_L	M_S	number of states	$L=2$ $S=1/2$	$L=1$ $S=1/2$	$L=0$ $S=3/2$
$1\alpha\, 0\alpha\, 1\beta$	2	$\frac{1}{2}$	1	✓		
$1\beta\, 0\beta\, 1\alpha$	2	$-\frac{1}{2}$	1	✓		
$1\alpha\, 0\alpha\, 0\beta$ $1\alpha\, \bar{1}\alpha\, 1\beta$	1	$\frac{1}{2}$	2	✓	✓	
$1\beta\, 0\beta\, 0\alpha$ $1\beta\, \bar{1}\beta\, 1\alpha$	1	$-\frac{1}{2}$	2	✓	✓	
$1\alpha\, 0\alpha\, \bar{1}\alpha$	0	$\frac{3}{2}$	1			✓
$1\alpha\, 0\alpha\, \bar{1}\beta$ $1\alpha\, \bar{1}\alpha\, 0\beta$ $0\alpha\, \bar{1}\alpha\, 1\beta$	0	$\frac{1}{2}$	3	✓	✓	✓
$1\beta\, 0\beta\, \bar{1}\alpha$ $1\beta\, \bar{1}\beta\, 0\alpha$ $0\beta\, \bar{1}\beta\, 1\alpha$	0	$-\frac{1}{2}$	3	✓	✓	✓
$1\beta\, 0\beta\, \bar{1}\beta$	0	$-\frac{3}{2}$	1			✓
$\bar{1}\alpha\, 0\alpha\, 0\beta$ $\bar{1}\alpha\, 1\alpha\, \bar{1}\beta$	-1	$\frac{1}{2}$	2	✓	✓	
$\bar{1}\beta\, 0\beta\, 0\alpha$ $\bar{1}\beta\, 1\beta\, \bar{1}\alpha$	-1	$-\frac{1}{2}$	2	✓	✓	
$\bar{1}\alpha\, 0\alpha\, \bar{1}\beta$	-2	$\frac{1}{2}$	1	✓		
$\bar{1}\beta\, 0\beta\, \bar{1}\alpha$	-2	$-\frac{1}{2}$	1	✓		
		total	20			

For each product, the M_L value is the sum of the three m_l values; similarly the M_S value is the sum of the three m_s values. Thus for the first product function, $M_L = 2$, and $M_S = \frac{1}{2}$. The M_L and M_S values for each of the product states are given in the second and third columns of the table.

(c) There are 20 states characterised by the set of quantum numbers L, S, M_L, M_S. These states are linear combinations of the 20 product states, and they must be antisymmetric for the interchange of any pair of electrons. For the case of two electrons in the same shell – Problem 7.6(c) – the antisymmetry requirement is satisfied by combining a symmetric space function with an antisymmetric spin function, and vice-versa. However, when there are more than two electrons in the valency shell, the problem is somewhat more complicated, because it is not, in general, possible to express the state function as the product of a space function and a spin function. Instead we construct an antisymmetric function for each product function by the method of (7.9). (Note that the M_L, M_S values of the antisymmetrised function are the same as those of the initial product from which it is constructed.) Each L, S, M_L, M_S state is a linear combination of the antisymmetrised product functions with the same values of M_L, M_S. This relationship determines the possible values of the L, S combinations.

The first line in Table 7.3 gives the values $M_L = 2$, $M_S = \frac{1}{2}$. This is the largest value of M_L, so the largest value of L is 2 (since $M_L \leqslant L$), and it is associated with $S = \frac{1}{2}$. The L, S, M_L, M_S state with the values 2, $\frac{1}{2}$, 2, $\frac{1}{2}$ is equal to the antisymmetrised product function 1α, 0α, 1β. The pair of values $L = 2$, $S = \frac{1}{2}$ corresponds to ten M_L, M_S combinations ($M_L = 2$, 1, 0, -1, -2, each of which may be combined with one of the two values $M_S = \frac{1}{2}$, $-\frac{1}{2}$.) Each combination is represented by a tick in the column $L = 2$, $S = \frac{1}{2}$ and the appropriate M_L, M_S row in the table.

Two product states appear in the third line of the table, with the values $M_L = 1$, $M_S = \frac{1}{2}$. So $L = 1$, $S = \frac{1}{2}$ is another pair of L, S values, giving rise to six M_L, M_S combinations shown by ticks in the table. The state with the L, S, M_L, M_S values 2, $\frac{1}{2}$, 1, $\frac{1}{2}$ is one linear combination of the two antisymmetrised product functions, and the state 1, $\frac{1}{2}$, 1, $\frac{1}{2}$ is another linear combination of the same two functions. Finally, the fifth line of the table with $M_L = 0$, $M_S = \frac{3}{2}$ shows that $L = 0$, $S = \frac{3}{2}$ is a third pair of L, S values, with the four M_L, M_S combinations indicated in the table. We thus have a total of $10 + 6 + 4 = 20$ states, which is the number required.

Each pair of L, S values gives rise to values of J from $L + S$, decreasing by unity to $|L - S|$. The L, S, J values are listed in Table 7.4. The states are thus

$${}^2D_{5/2},\ {}^2D_{3/2},\ {}^2P_{3/2},\ {}^2P_{1/2},\ {}^4S_{3/2}.$$

Table 7.4. *L, S, J values and number of states for the electronic configuration $(np)^3$.*

L	S	J	$2J+1$	sum of $2J+1$
2	$\frac{1}{2}$	$\frac{5}{2}, \frac{3}{2}$	6, 4	10
1	$\frac{1}{2}$	$\frac{3}{2}, \frac{1}{2}$	4, 2	6
0	$\frac{3}{2}$	$\frac{3}{2}$	4	4
			total	20

(d) The state with the lowest energy is given by Hund's rules, p. 131. The first rule gives $S=\frac{3}{2}$. So the ground state is $^4S_{3/2}$.

7.8 Since lead obeys *jj* coupling, we need the j values for a single electron, which, for an electron in a p shell, are $\frac{3}{2}$ and $\frac{1}{2}$. For $j_1 = j_2 = \frac{3}{2}$, $J = 3, 2, 1, 0$. However, the values 3 and 1 correspond to symmetric functions (p. 129) and are therefore rejected. Similarly, for $j_1 = j_2 = \frac{1}{2}$, $J = 1, 0$; again $J = 1$ is symmetric and is rejected. For $j_1 = \frac{3}{2}$, $j_2 = \frac{1}{2}$, $J = 2$, 1. Since the two values of j are different we can form antisymmetric functions for both values of J. The permitted values of j_1, j_2, J are listed in Table 7.5.

Table 7.5. *j_1, j_2, J values and number of states for the electronic configuration $(np)^2$.*

j_1	j_2	J	$2J+1$	sum of $2J+1$
$\frac{3}{2}$	$\frac{3}{2}$	2, 0	5, 1	6
$\frac{3}{2}$	$\frac{1}{2}$	2, 1	5, 3	8
$\frac{1}{2}$	$\frac{1}{2}$	0	1	1
			total	15

Comment

For the electronic configuration $(np)^2$, the results of Problem 7.6(c) give the L, S, J values for LS coupling, and the present results give the j_1, j_2, J values for *jj* coupling. In Fig. 7.1 we show schematically how the energies of the states of the $(np)^2$ configuration vary with P/V', the ratio of the magnetic and residual electrostatic forces in the atom – see p. 131. For $P \ll V'$, we have LS coupling, exemplified by germanium. Inter-

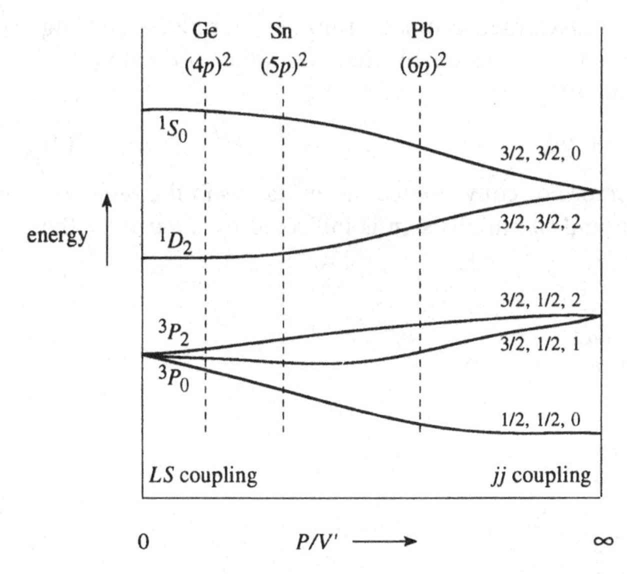

Fig. 7.1. *Schematic diagram showing the form of the energy levels for the $(np)^2$ configuration as a function of the ratio of P, the magnetic interactions, to V', the residual electrostatic interactions in the atom. $P/V' \ll 1$ corresponds to LS coupling, and $P/V' \gg 1$ corresponds to jj coupling. The L, S, J values and the j_1, j_2, J values are shown for the two limiting cases. The diagram is adapted from Condon and Shortley (1963).*

mediate coupling, when $P \sim V'$, is shown by tin. Here the states are, in general, mixtures of states with different L, S values, and also of states with different j_1, j_2 values. For $P \gg V'$, the jj coupling condition, the states remain mixtures of L, S values, but tend to single terms in j_1, j_2. This is the case for lead. Note that J is a good quantum number for all values of V' and P.

7.9 The possible values of the spin of the nucleus are the J values obtained from the addition of the angular momentum of the three neutrons with $j_1 = j_2 = j_3 = \frac{5}{2}$. The neutron has spin $\frac{1}{2}$, so the state function of the three neutrons must be antisymmetric. For $j = \frac{5}{2}$, there are $2j + 1 = 6$ values of m_j. The m_j values for a product state of three neutrons must be different. Therefore the number of product states is equal to the number of ways of selecting three different m_j values from six, which is

$$\frac{6!}{3!3!} = 20. \tag{1}$$

From each product an antisymmetric function may be formed according to (7.9). The product states are listed by their m_j values in Table 7.6, those with the same value of

$$M_J = m_{j_1} + m_{j_2} + m_{j_3} \tag{2}$$

being put on the same line. For convenience the m_j values in the table are twice the actual values, and the minus sign is indicated by a bar over the number.

Table 7.6. *Product states and M_J values for three fermions with $j = \frac{5}{2}$.*

product states			M_J
$5\,3\,1$			9/2
$5\,3\,\bar{1}$			7/2
$5\,3\,\bar{3}$	$5\,1\,\bar{1}$		5/2
$5\,3\,\bar{5}$	$5\,1\,\bar{3}$	$3\,1\,\bar{1}$	3/2
$5\,1\,\bar{5}$	$5\,\bar{1}\,\bar{3}$	$3\,1\,\bar{3}$	1/2
$\bar{5}\,\bar{1}\,5$	$\bar{5}\,1\,3$	$\bar{3}\,\bar{1}\,3$	−1/2
$\bar{5}\,\bar{3}\,5$	$\bar{5}\,\bar{1}\,3$	$\bar{3}\,\bar{1}\,1$	−3/2
$\bar{5}\,\bar{3}\,3$	$\bar{5}\,\bar{1}\,1$		−5/2
$\bar{5}\,\bar{3}\,1$			−7/2
$\bar{5}\,\bar{3}\,\bar{1}$			−9/2

Each state $\Psi(J, M_J)$ is a linear combination of the antisymmetrised product states with the same value of M_J. Conversely, each antisymmetrised product is a linear combination of the states $\Psi(J, M_J)$ with the same M_J and different J. It follows therefore, since $|M_J| \leqslant J$, that if there are n product functions with the same value of M_J, there are n values of J equal to, or greater than, M_J. Inspection of the table shows that the required values of J are $\frac{9}{2}$, $\frac{5}{2}$, $\frac{3}{2}$. These three values give $10 + 6 + 4 = 20$ states as expected.

Comment

The three J values are the possible values of the spin of the ground state of the nucleus ^{19}O according to the shell model. In a further version of the model, known as the *independent-particle model*, the spins of pairs of nucleons in unfilled shells couple to give zero. Thus the spin of the nucleus is zero if the number of nucleons is even, and equal to the spin of the remaining nucleon if the number is odd. So the model predicts that the spin of the ground state of ^{19}O is $J = \frac{5}{2}$, which is the correct value.

7.10 (a) The nuclei in the light hydrogen molecule are protons, which have spin $\frac{1}{2}$, so the overall state function of the two nuclei is antisymmetric. The rotational eigenstates are the spherical harmonics Y_{lm}. They are symmetric or antisymmetric with respect to the interchange of the two nuclei according to whether l is even or odd – see the solution to Problem 7.4. The spin state for the pair of protons is symmetric for the total spin $S = 1$ (ortho molecule), and antisymmetric for $S = 0$ (para molecule). So, because the overall state function of the nuclei is antisymmetric, the para molecules have rotational states with even l, and the ortho molecules rotational states with odd l.

In an equilibrium mixture at temperature T, the number of molecules in each state is proportional to the Boltzmann factor $\exp(-E_l/k_B T)$, where the energy of the state is

$$E_l = \frac{\hbar^2}{2I} l(l+1). \tag{1}$$

Since the rotational state is the spherical harmonic Y_{lm}, each rotational energy level has a $(2l+1)$-fold space degeneracy, together with a three-fold spin degeneracy for those energies associated with the spin $S = 1$. From these considerations we arrive at the required result that the ratio of the number of para to the number of ortho molecules in an equilibrium mixture of light hydrogen at temperature T is

$$\frac{n_p}{n_o} = \frac{1}{3} \frac{\sum_{l\ \mathrm{even}} (2l+1) \exp\{-l(l+1)x\}}{\sum_{l\ \mathrm{odd}} (2l+1) \exp\{-l(l+1)x\}}, \tag{2}$$

where

$$x = \frac{\hbar^2}{2I} \frac{1}{k_B T}. \tag{3}$$

(b) If R_0 is the distance between the nuclei, and m_p is the mass of the proton, then

$$I = 2m_p \left(\frac{R_0}{2}\right)^2 = \tfrac{1}{2} m_p R_0^2. \tag{4}$$

The Boltzmann factor causes the terms to diminish very rapidly as l increases, and a rough estimate of R_0 shows that, at $T = 20$ K, terms with

$l > 1$ are negligible. We can see this, because we expect R_0 to be somewhere between a_0 and $2a_0$, where $a_0 = 53$ pm is the Bohr radius. Putting $R_0 = 100$ pm, together with $T = 20$ K, and the values of the atomic constants on p. 1, gives $x = 2.41$. So, for $l = 2$,

$$(2l + 1) \exp\{-l(l + 1)x\} = 5 \exp(-6x) = 2.6 \times 10^{-6}. \tag{5}$$

Retaining only the terms $l = 0, 1$ in (2), we have

$$\frac{n_p}{n_o} = \frac{1}{9 \exp(-2x)} = \frac{99.83}{0.17}, \tag{6}$$

which gives

$$x = 4.286. \tag{7}$$

Inserting this value in (3) and (4) we obtain

$$R_0 = 75 \text{ pm}. \tag{8}$$

(c) The nuclei in heavy hydrogen are deuterons with spin 1. So the ratio of symmetric spin states (ortho) to antisymmetric spin states (para) is 2:1 (Problem 7.1). The deuteron is a boson, so, the overall state function for the two nuclei is symmetric. The ortho states therefore have l even, and the para states have l odd. The ratio of the number of ortho to para molecules is thus

$$\frac{n_o}{n_p} = \frac{2 \sum\limits_{l \text{ even}} (2l + l) \exp\{-l(l + 1)x_d\}}{\sum\limits_{l \text{ odd}} (2l + 1) \exp\{-l(l + 1)x_d\}}, \tag{9}$$

where

$$x_d = \frac{\hbar^2}{m_d R_0^2 k_B T} = \tfrac{1}{2}x. \tag{10}$$

(m_d, the mass of the deuteron, is approximately $2m_p$.) As before, the terms with $l > 1$ are negligible. So

$$\frac{n_o}{n_p} = \frac{2}{3 \exp(-x)} = \frac{2}{3} \exp(4.286) = 48.5, \tag{11}$$

which corresponds to 98.0% ortho molecules.

Comment

In addition to the rotational motion of the hydrogen molecule, the distance between the two hydrogen atoms may change, giving rise to vibrational states. However, the spacing between the energy levels of these states is much larger than that of the rotational states, and at a temperature of 20 K the molecules are all in the lowest vibrational state. The energies of the states in which the electrons are excited are higher still. The vibrational and electronic ground states are symmetric with respect to the interchange of the two nuclei. Therefore it is correct to consider only the rotational and the nuclear spin states in the problem.

7.11 The two lowest rotational energy levels of the HD molecule correspond to $l = 0$ and 1, with degeneracies 1 and 3, as in the last problem. The spins of the two nuclei are

$$s_1 = \tfrac{1}{2}, \quad \text{and} \quad s_2 = 1. \tag{1}$$

Thus there are a total of $2 \times 3 = 6$ spin states. As the two particles are not identical, the state function has no symmetry. Any spin state may be combined with any space state. So the degeneracies of the $l = 0$ and $l = 1$ energy values are 6 and 18.

7.12 (a) The values of L, S, J for the ground state of the praseodymium ion are given by Hund's rules, p. 131. For two electrons in the $4f$ shell, the rules give

$$S = 1,\ L = 5,\ J = 4. \tag{1}$$

The g-factor for the ion is

$$g = 1 + \frac{J(J+1) + S(S+1) - L(L+1)}{2J(J+1)} = \frac{4}{5}. \tag{2}$$

(b) In the presence of a magnetic field **B**, the component of magnetic dipole moment of an ion in the direction of **B** is

$$\mu = gM_J\mu_B, \tag{3}$$

and its magnetic energy is

$$E = -gM_J\mu_B B, \tag{4}$$

where M_J takes values $-J,\ -J+1,\ -J+2,\ \ldots,\ J$. The probability p that an ion has a particular value M_J is proportional to the Boltzmann factor $\exp(-E/k_B T)$. Thus

$$p = \frac{\exp(M_J g\mu_B B/k_B T)}{Z}, \tag{5}$$

where

$$Z = \sum_{M_J=-J}^{M_J=J} \exp(M_J g\mu_B B/k_B T). \tag{6}$$

Note that if we keep J constant and sum p over M_J, the result is unity. This is because the expressions for p in (5) and (6) are based on the assumption that the states of larger J that can be formed from the same values of L and S are so much higher in energy that, at the temperature of the problem, the probability of the ion being in one of these states is negligible. This is what is meant by the statement at the end of the problem that $k_B T$ is small compared to the LS multiplet spacing.

From (3) to (6), the thermal average of μ is

$$\begin{aligned}\bar{\mu} &= \left\{\sum_{M_J=-J}^{M_J=J} M_J g\mu_B \exp(M_J g\mu_B B/k_B T)\right\}/Z \\ &= g\mu_B\left\{\sum_{M_J=-J}^{M_J=J} M_J \exp(M_J x)\right\}\bigg/\left\{\sum_{M_J=-J}^{M_J=J} \exp(M_J x)\right\},\end{aligned} \tag{7}$$

where

$$x = g\mu_B B/k_B T. \tag{8}$$

Note that x remains constant in the sum over M_J. For $B = 1\,\text{T}$, and $T = 300\,\text{K}$,

$$\frac{\mu_B B}{k_B T} = 0.0022. \tag{9}$$

Therefore, since g and M_J are both of the order of unity, $M_J x \ll 1$, and we may put

$$\exp(M_J x) \approx 1 + M_J x, \tag{10}$$

which gives

$$\bar{\mu} = g\mu_B \frac{\sum(M_J + xM_J^2)}{\sum(1 + xM_J)}. \tag{11}$$

The summation over M_J is from $-J$ to $+J$, so the sum over an odd power of M_J is zero. The non-zero terms are

$$\sum M_J^2 = \tfrac{1}{3}J(J+1)(2J+1), \tag{12}$$

and

$$\sum 1 = 2J + 1. \tag{13}$$

Thus

$$\bar{\mu} = \tfrac{1}{3}g^2\mu_B^2\frac{J(J+1)B}{k_B T}. \tag{14}$$

(c) The magnetic susceptibility per mole is defined by

$$\chi_{mol} = \frac{\mu_0 N_A \bar{\mu}}{B}, \tag{15}$$

where N_A is the Avogadro constant. So from (14)

$$\chi_{mol} = \tfrac{1}{3}\mu_0 N_A g^2 \mu_B^2 \frac{J(J+1)}{k_B T}. \tag{16}$$

Substituting the values of g, J, T, and those of the atomic constants, gives

$$\chi_{mol} = 6.7 \times 10^{-8}\ \mathrm{m^3\ mol^{-1}}. \tag{17}$$

Comments

(1) It can be seen from (16), that, provided $\mu_B B \ll k_B T$, the magnetic susceptibility is independent of B, and is inversely proportional to the temperature. This latter variation is known as *Curie's law*. The condition $\mu_B B \ll k_B T$ is satisfied for a large range of experimental conditions, so the law has a wide validity.

(2) An LS multiplet is said to be *wide* if the energy separation of the states of different J is large compared to $k_B T$, where T is of the order of room temperature. This is the case for the ions of the rare-earth elements such as praseodymium. Provided the condition for Curie's law is satisfied, the experimental values of the magnetic susceptibility of rare-earth ions agree well with the theoretical expression in (16).

The iron group of ions have *narrow* LS multiplets, i.e. the energy separation of the states of different J is small compared to the room-temperature value of $k_B T$. A narrow LS multiplet means that the spin–orbit forces in the atom are relatively weak – see p. 131. We may therefore ignore the coupling of the orbital and spin angular momenta, and regard the atom as having independent orbital and spin magnetic dipole moments, each of which contributes to the magnetic susceptibility. The values of g are 1 for orbital, and 2 for spin motion. Thus, for a narrow multiplet, (16) becomes

$$\chi_{mol} = \tfrac{1}{3}\mu_0 N_A \mu_B^2 \frac{L(L+1) + 4S(S+1)}{k_B T}. \tag{18}$$

However, the values derived from this expression do not agree with those of experiment. The reason is that, in salts of the iron group, the electrostatic interatomic forces in the crystal cause the value of L to be effectively zero, a phenomenon known as *quenching*. The values obtained from (18), with $L = 0$, are found to be in good agreement with experiment.

8

Time, time-dependent perturbation theory, transitions

Summary of theory

1 Time-varying Hamiltonian

(a) General problem

Finding the eigenfunctions and eigenvalues of a general time-varying Hamiltonian is difficult, but the problem is greatly simplified if the Hamiltonian has the form

$$H = H^{(0)} + H^{(1)}, \tag{1}$$

where $H^{(0)}$ does not vary with time, and $H^{(1)}$ ($\ll H^{(0)}$) varies with time. It is assumed that the eigenfunctions u_j and eigenvalues E_j of $H^{(0)}$ are known for all j. The state function at time t can be expressed as

$$\psi(t) = \sum_j c_j(t) \exp(-\mathrm{i}E_j t/\hbar) u_j. \tag{2}$$

We suppose that all the coefficients $c_j(0)$ are known. The problem is to calculate $c_j(t)$ as $\psi(t)$ varies according to the 6th postulate

$$\frac{\partial \psi}{\partial t} = \frac{1}{\mathrm{i}\hbar} H \psi. \tag{3}$$

If we substitute (2) into (3) and use the orthonormal properties of the u_j, we arrive at the result

$$\dot{c}_j(t) = \frac{\partial c_j(t)}{\partial t} = \frac{1}{\mathrm{i}\hbar} \sum_n c_n(t) \langle j | H^{(1)}(t) | n \rangle \exp(\mathrm{i}\omega_{jn} t), \tag{4}$$

where

$$\langle j | H^{(1)}(t) | n \rangle = \int_{\substack{\text{all}\\ \text{space}}} u_j^* H^{(1)}(t) u_n \, \mathrm{d}\tau, \tag{5}$$

and

$$\hbar\omega_{jn} = E_j - E_n. \tag{6}$$

(b) First-order time-dependent perturbation theory

The result in (4) is exact, but is, in general, not useful, because the right-hand side contains the time-varying coefficients $c_n(t)$, which are the quantities we are trying to find. We therefore approximate by replacing the $c_n(t)$ by their values at $t = 0$. This is equivalent to first-order perturbation theory. Further, for simplicity, we consider the case where, at $t = 0$, the system is in an eigenstate of $H^{(0)}$ – denote it by k. Thus

$$c_k(0) = 1, \quad c_n(0) = 0, \quad n \neq k. \tag{7}$$

There is now only one term, $n = k$, on the right-hand side of (4), which becomes

$$\dot{c}_j(t) = \frac{1}{\mathrm{i}\hbar}\langle j|H^{(1)}(t)|k\rangle \exp(\mathrm{i}\omega_{jk}t), \quad j \neq k, t > 0, \tag{8}$$

whence

$$c_j(t) = \frac{1}{\mathrm{i}\hbar}\int_0^t \langle j|H^{(1)}(t')|k\rangle \exp(\mathrm{i}\omega_{jk}t')\,\mathrm{d}t'. \tag{9}$$

In this approximation, $c_k(t) = 1$. It is assumed that E_k is non-degenerate. The procedure for the degenerate case is similar to that for time-independent perturbation theory and will not be considered here.

2 Einstein *A* and *B* coefficients

(a) Definitions

Consider a set of atoms with two states k and j, having energies E_k and E_j, where $E_k > E_j$. The Einstein coefficient A is the probability per unit time that an atom in state k makes a spontaneous transition to state j. The coefficient B is defined for transitions stimulated by electromagnetic radiation. If the energy density of the radiation is $I(\omega)\,\mathrm{d}\omega$ in the angular frequency range ω to $\omega + \mathrm{d}\omega$, then the probability per unit time of a transition between the two states, in either direction, is $BI(\omega_{kj})$, where $\hbar\omega_{kj} = E_k - E_j$.

(b) Electric dipole transitions

For electric dipole transitions the coefficients are given by

$$A = \frac{4}{3}\frac{e^2}{4\pi\varepsilon_0}\frac{1}{\hbar c^3}\omega_{kj}^3 |\langle j|\mathbf{r}|k\rangle|^2, \tag{10}$$

$$B = \frac{4\pi^2}{3} \frac{e^2}{4\pi\varepsilon_0} \frac{1}{\hbar^2} |\langle j|\mathbf{r}|k\rangle|^2, \tag{11}$$

where **r** is the position operator.

3 Planck law for the spectral distribution of black-body radiation

The energy density function is

$$I(\omega) = \frac{\hbar\omega^3}{\pi^2 c^3} \frac{1}{\exp(\hbar\omega/k_B T) - 1}. \tag{12}$$

Problems

8.1 A beam of thermal neutrons, polarised in the x direction, passes at time $t = 0$ into a region of constant magnetic field **B** in the z direction.

(a) Show that the spin state of the neutrons $\psi(t)$ is an eigenstate of the operator S_ρ, corresponding to the component of spin angular momentum in the direction $(\cos\rho, \sin\rho, 0)$, where $\rho = 2\mu_n Bt/\hbar$. (μ_n is the magnetic moment of the neutron.)

(b) Show that this result is in accord with the classical result for a precessing gyroscope.

(c) For $B = 10.00$ mT, the time for a 2π rotation of the spin is 3.429 μs. Calculate the value of μ_n in nuclear magnetons.

8.2 (a) A beam of orthohydrogen molecules (proton spin state $S = 1$), travelling along the y axis, passes through a Stern–Gerlach apparatus with its magnetic field along the x axis – Fig. 8.1. The emerging molecules with $M_S = 1$ are passed through a second Stern–Gerlach apparatus with its magnetic field along the x axis. A constant magnetic field **B** in the z direction acts along part of the path between the magnets. Show that the fraction of molecules emerging from the three output channels, corresponding to $M_S = 1, 0, -1$, of the second magnet is

$$\cos^4(\omega t/2), \quad 2\sin^2(\omega t/2)\cos^2(\omega t/2), \quad \sin^4(\omega t/2),$$

where $\omega = 2\mu_p B/\hbar$, and t is the time the molecules spend in the field **B**. (μ_p is the magnetic moment of the proton.)

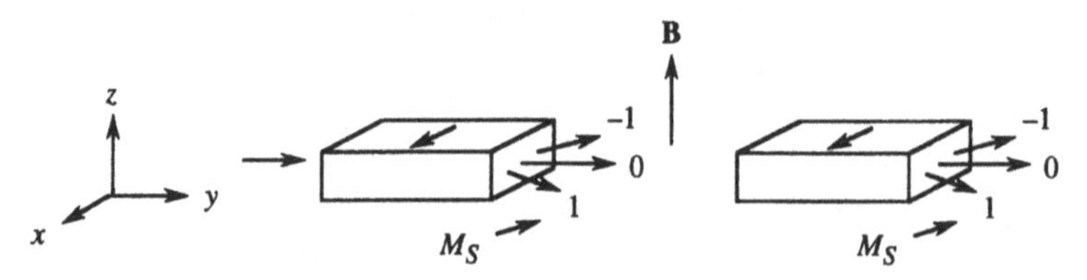

Fig. 8.1. *Geometry for Problem 8.2.*

(b) For a path length in **B** of 20 mm, and molecules with energy 25 meV, it is found that no molecules emerge from the $M_S = 1$ channel of the second magnet when

$$B = 1.80\,(n + \tfrac{1}{2})\,\text{mT},$$

where n is an integer. Deduce the value of μ_p in nuclear magnetons. [The spins of the electrons in orthohydrogen are antiparallel.]

8.3 A one-dimensional harmonic oscillator has mass m and angular frequency ω. A time-dependent state $\psi(t)$ of the oscillator is given at $t = 0$ by

$$\psi(0) = \frac{1}{\sqrt{(2s)}} \sum |n\rangle,$$

where $|n\rangle$ is an eigenstate of the Hamiltonian corresponding to the quantum number n, and the summation runs from $n = N - s$ to $n = N + s$, with $N \gg s \gg 1$.

(a) Show that the expectation value of the displacement varies sinusoidally with amplitude $(2\hbar N/m\omega)^{1/2}$.

(b) Relate this result to the time variation of the displacement of a classical harmonic oscillator.

8.4 A system of hydrogen atoms in the ground state is contained between the plates of a parallel-plate capacitor. A voltage pulse is applied to the capacitor so as to produce a homogeneous electric field

$$\mathscr{E} = 0 \quad t < 0, \quad \mathscr{E} = \mathscr{E}_0 \exp(-t/\tau) \quad t > 0.$$

(a) Show that, after a long time, the fraction of atoms in the $2p$ $(m = 0)$ state is, to first order,

$$\frac{2^{15}}{3^{10}} \frac{a_0^2 e^2 \mathscr{E}_0^2}{\hbar^2(\omega^2 + 1/\tau^2)},$$

where a_0 is the Bohr radius, and $\hbar\omega$ is the energy difference between the $2p$ and the ground state.

(b) What is the fraction of atoms in the $2s$ state?

8.5 A system of atoms can make radiative transitions from an excited state to the ground state. If the probability per unit time of a transition is γ, show that the power spectrum of the radiation is a Lorentzian whose angular frequency width at half-height is equal to γ.

8.6 A time-varying Hamiltonian $H^{(1)}(t')$ brings about transitions of a system from a state k at $t' = 0$ to a state j at $t' = t$ with probability $p_{k\rightarrow j}(t)$. Use first-order time-dependent perturbation theory to show that, if $p_{j\rightarrow k}(t)$ is the probability that the same Hamiltonian brings about the transition $j \rightarrow k$ in the same time interval, then

$$p_{j\rightarrow k}(t) = p_{k\rightarrow j}(t).$$

8.7 A set of identical atoms, which have two states k and j with non-degenerate energies E_k and E_j ($E_k > E_j$), are in a box whose walls are at a constant temperature. By considering the equilibrium numbers of atoms in the two states, show that

$$\frac{A}{B} = \frac{\hbar\omega_{kj}^3}{\pi^2 c^3},$$

where A and B are the Einstein coefficients for spontaneous and stimulated transitions, and $\hbar\omega_{kj} = E_k - E_j$.

8.8 Starting from the expression (8.10) for the Einstein coefficient for spontaneous emission, show that, for an atomic electric dipole transition to the ground state in hydrogen, the fractional frequency width is of the order of α^3, where $\alpha = e^2/4\pi\varepsilon_0 c\hbar$ is the fine structure constant.

8.9 A mercury lamp emits radiation of wavelength 254 nm, with a fractional wavelength spread of 10^{-5}. If the output flux is $1\,\mathrm{kW\,m^{-2}}$, estimate the ratio of stimulated to spontaneous emission processes in the lamp.

8.10 A superconducting tunnelling junction consists of two identical superconductors 1 and 2, separated by a thin insulating layer – Fig. 8.2. The superconducting electron pairs in each superconductor are in the same quantum state with a single phase. If n_1 denotes the density of these pairs in 1, their wavefunction ψ_1 may be expressed in the form

$$\psi_1 = n_1^{1/2} \exp(\mathrm{i}\theta_1),$$

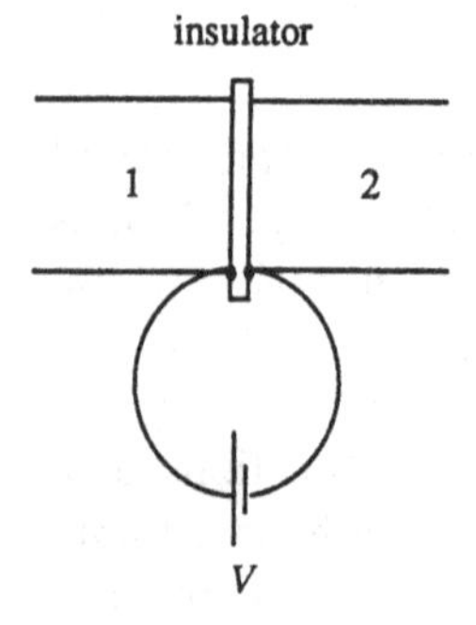

Fig. 8.2. *Josephson tunnelling junction. 1 and 2 are superconductors.*

where θ_1 is the common phase angle. Similarly

$$\psi_2 = n_2^{1/2} \exp(i\theta_2).$$

The ψs, ns, and θs are all time-dependent. The time variation of ψ_1 is given by

$$\frac{\partial \psi_1}{\partial t} = \frac{1}{i\hbar}(E_1 \psi_1 + F\psi_2),$$

where E_1 is the energy of the superconducting electrons in 1, and the term $F\psi_2$, where F is a real constant, represents electron pairs tunnelling from 2 to 1. Similarly

$$\frac{\partial \psi_2}{\partial t} = \frac{1}{i\hbar}(E_2 \psi_2 + F\psi_1).$$

(a) Show that

$$\frac{\partial n_1}{\partial t} = 2\Omega n \sin(\theta_2 - \theta_1),$$

where $\Omega = F/\hbar$, and $n \approx n_1 \approx n_2$.

(b) Hence show that if a steady voltage V is applied across the junction, the current of superconducting electrons oscillates with frequency $\nu = 2eV/h$.

Solutions

8.1 (a) The neutron, like the electron, has spin $\frac{1}{2}$. Thus all the previous results for the spin angular momentum operators of the electron apply. (We shall be concerned only with spin angular momentum in the present problem and shall therefore drop the word 'spin'.) The operator S_z has eigenfunctions α and β, with eigenvalues $\pm\frac{1}{2}\hbar$. The magnetic moment μ_n of the neutron (which we take as a positive quantity) refers to the component value. Its direction is opposite to that of the angular momentum vector, i.e. the magnetic moment of the neutron is to be regarded as due to the rotation of a negative charge distribution.

At time $t = 0$, the state function is the eigenfunction of S_x, the operator for the x component of angular momentum, with eigenvalue $+\frac{1}{2}\hbar$, i.e.

$$\psi(0) = (\alpha + \beta)/\sqrt{2} \tag{1}$$

– see Problem 5.3. In the region where the magnetic field acts, the Hamiltonian has a term proportional to S_z, so the eigenfunctions of the Hamiltonian are α and β. The form of $\psi(0)$ in (1) is therefore the one we require to calculate $\psi(t)$. For the states α and β, the energies are

$$E_\alpha = \mu_n B, \quad E_\beta = -\mu_n B. \tag{2}$$

Thus from (2.12)

$$\psi(t) = \{\alpha \exp(-i\omega t) + \beta \exp(i\omega t)\}/\sqrt{2}, \tag{3}$$

where

$$\omega = \mu_n B/\hbar. \tag{4}$$

Now consider the direction $(\cos\rho, \sin\rho, 0)$, i.e. the direction in the xy plane making an angle ρ with the x axis. The Pauli spin operator for this direction is

$$\sigma_\rho = \cos\rho\,\sigma_x + \sin\rho\,\sigma_y. \tag{5}$$

Using the results – see (5.3.6) –

$$\sigma_x \alpha = \beta, \quad \sigma_y \alpha = i\beta, \tag{6}$$

we obtain

$$\begin{aligned}\sigma_\rho \alpha &= \cos\rho\,\sigma_x \alpha + \sin\rho\,\sigma_y \alpha \\ &= (\cos\rho + i\sin\rho)\beta = \exp(i\rho)\beta.\end{aligned} \tag{7}$$

Similarly

$$\sigma_\rho \beta = \exp(-i\rho)\alpha. \tag{8}$$

Thus

$$\sigma_\rho\{\exp(-i\rho/2)\alpha + \exp(i\rho/2)\beta\}/\sqrt{2}$$
$$= \{\exp(i\rho/2)\beta + \exp(-i\rho/2)\alpha\}/\sqrt{2}, \tag{9}$$

i.e. the function

$$\{\exp(-i\rho/2)\alpha + \exp(i\rho/2)\beta\}/\sqrt{2} \tag{10}$$

is an eigenfunction of σ_ρ with eigenvalue 1. Comparison of (3) and (10) shows that $\psi(t)$ is an eigenfunction of σ_ρ with eigenvalue 1, where

$$\rho = 2\omega t = \frac{2\mu_n Bt}{\hbar}. \tag{11}$$

The quantum mechanical result is thus that the spin precesses in the xy plane in the direction $x \rightarrow y$ with angular velocity

$$\omega_L = 2\omega = \frac{2\mu_n B}{\hbar}. \tag{12}$$

(b) The geometry for the classical description of the motion is shown in Fig. 8.3. The angular momentum vector is along $+x$, and the magnetic moment vector along $-x$. The magnetic field gives a couple

$$\mathbf{G} = \boldsymbol{\mu}_n \times \mathbf{B} = \mu_n B, \tag{13}$$

in the y direction. In time δt the couple creates angular momentum $G\delta t$ along y. Since the neutron possesses angular momentum $\frac{1}{2}\hbar$ along x, the required angular momentum along y is obtained if the angular momentum vector moves through an angle $\delta\theta$ as shown, where

$$\tfrac{1}{2}\hbar \sin\delta\theta \approx \tfrac{1}{2}\hbar\delta\theta = \mu_n B\delta t. \tag{14}$$

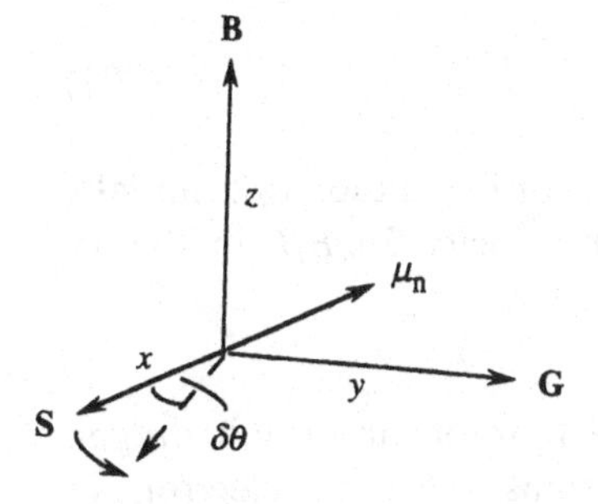

Fig. 8.3. *Precession of neutron spin in a magnetic field.*

Thus the angular velocity of precession is

$$\omega_L = \frac{\delta\theta}{\delta t} = \frac{2\mu_n B}{\hbar}, \tag{15}$$

which is the same result as in (12).

The quantity ω_L is known as the *Larmor frequency*. In general, a uniform magnetic field **B** causes the magnetic moment of a particle to precess around the direction of **B** with angular frequency

$$\omega_L = \gamma B, \tag{16}$$

where γ, known as the *gyromagnetic ratio*, is the ratio of the components of magnetic moment and angular momentum. The sense of the precession depends on whether the magnetic moment and angular momentum are in the same or opposite directions.

(c) If t_0 is the time for a 2π rotation of the spin

$$\omega_L = \frac{2\pi}{t_0} = \frac{2\mu_n B}{\hbar}, \tag{17}$$

i.e.

$$\mu_n = \frac{\pi\hbar}{Bt_0} \tag{18}$$

$$= 9.662 \times 10^{-27}\ \mathrm{J\,T^{-1}} = 1.913\ \mu_N. \tag{19}$$

Comment

The relation between the eigenfunctions of S_x and those of the Hamiltonian H, is the same as that for the operator A and H in Problem 2.7. So the result (2.7.7) applies. Putting

$$a_1 = \frac{\hbar}{2}, \quad a_2 = -\frac{\hbar}{2}, \quad E_1 = -E_2 = \mu_n B, \tag{20}$$

gives

$$\langle S_x \rangle = \frac{\hbar}{2}\cos\frac{2\mu_n Bt}{\hbar}, \tag{21}$$

which represents the x component of a vector of magnitude $\hbar/2$, initially along the x axis, and rotating with angular velocity $2\mu_n B/\hbar$ in the xy plane.

8.2 (a) The spin quantum number of the two protons in orthohydrogen is $S = 1$, with M_S values 1, 0, -1. Since the spins of the two electrons in the molecule are antiparallel, the fields in the Stern–Gerlach magnets, and in the region between them, act only on the magnetic moments of the protons. The spin state function ψ at $t = 0$ (defined to be the time when the molecule enters the magnetic field **B** along z) is an eigenfunction of S_x with eigenvalue $M_S = 1$. We need to express it in terms of the eigenfunctions of the Hamiltonian, which are the eigenfunctions of S_z.

We use the results of Problem 4.2. The latter was stated to relate to orbital angular momentum, but the essential feature is that $l = 1$, the same value as S here. So the relations between the eigenfunctions of the x and z component operators are identical. If ϕ_1, ϕ_0, ϕ_{-1} are the eigenfunctions of S_z, the eigenfunction of S_x with eigenvalue $M_S = 1$ is $(\phi_1 + \sqrt{2}\phi_0 + \phi_{-1})/2$, i.e.

$$\psi(0) = (\phi_1 + \sqrt{2}\phi_0 + \phi_{-1})/2. \tag{1}$$

The magnetic moment of the orthohydrogen molecule is $\mu = 2\mu_p$. So the energies for the states ϕ_1, ϕ_0, ϕ_{-1} in the magnetic field are

$$E_1 = -\mu B, \quad E_0 = 0, \quad E_{-1} = \mu B. \tag{2}$$

Thus

$$\psi(t) = \{\phi_1 \exp(\mathrm{i}\omega t) + \sqrt{2}\phi_0 + \phi_{-1}\exp(-\mathrm{i}\omega t)\}/2, \tag{3}$$

where

$$\omega = \frac{\mu B}{\hbar}. \tag{4}$$

We now express $\psi(t)$, where t is the time the molecules leave the field **B**, in terms of the eigenfunctions of S_x, i.e. we put

$$\psi(t) = c_1(\phi_1 + \sqrt{2}\phi_0 + \phi_{-1})/2 + c_0(\phi_1 - \phi_{-1})/\sqrt{2}$$
$$+ c_{-1}(\phi_1 - \sqrt{2}\phi_0 + \phi_{-1})/2, \tag{5}$$

again using the results of Problem 4.2. Then $|c_1|^2$, $|c_0|^2$, $|c_{-1}|^2$ give the relative numbers of molecules emerging from the three output channels of the second Stern–Gerlach magnet. The quantities c_1, c_0, c_{-1} are obtained by equating the coefficients of ϕ_1, ϕ_0, ϕ_{-1} in (3) and (5). Thus

$$c_1 + c_{-1} + \sqrt{2}c_0 = \exp(\mathrm{i}\omega t), \tag{6}$$

$$c_1 + c_{-1} - \sqrt{2}c_0 = \exp(-\mathrm{i}\omega t), \tag{7}$$

$$c_1 - c_{-1} = 1, \tag{8}$$

whence

$$c_1 = \cos^2\frac{\omega t}{2}, \quad c_{-1} = \sin^2\frac{\omega t}{2}, \quad c_0 = \frac{\mathrm{i}}{\sqrt{2}}\sin\omega t. \tag{9}$$

Thus

$$c_1^2 = \cos^4\frac{\omega t}{2}, \quad c_{-1}^2 = \sin^4\frac{\omega t}{2}, \quad |c_0|^2 = 2\sin^2\frac{\omega t}{2}\cos^2\frac{\omega t}{2}. \tag{10}$$

The quantities $|c_1|^2$, $|c_0|^2$, $|c_{-1}|^2$ are plotted against ωt in Fig. 8.4.

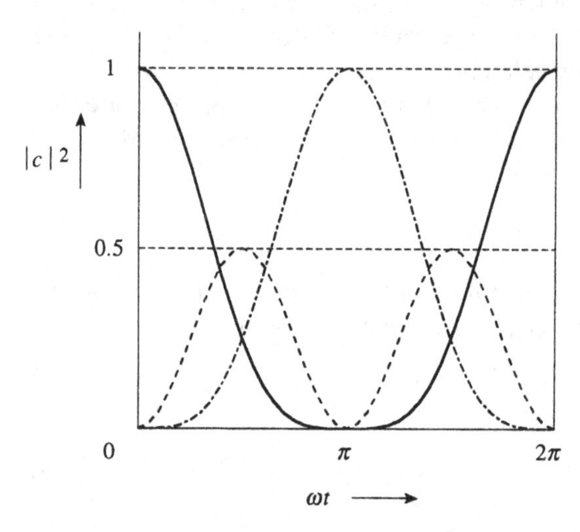

Fig. 8.4. *Problem 8.2 – variation of probability coefficients with time:* —— $|c_1|^2$, ---- $|c_{-1}|^2$, ---- $|c_0|^2$.

(b) No molecules emerge from the $M_S = 1$ channel of the second Stern–Gerlach magnet when $|c_1|^2 = 0$, which, from (9), is when

$$\omega t = (2n + 1)\pi, \tag{11}$$

where $n = 0, 1, 2, \ldots$. From (4) this corresponds to

$$\frac{\mu B}{\hbar}t = \frac{2\mu_p B}{\hbar}t = (2n + 1)\pi, \tag{12}$$

i.e.

$$\mu_p = \frac{\pi\hbar}{B_0 t}, \tag{13}$$

where

$$B_0 = 1.80 \text{ mT}. \tag{14}$$

If E is the kinetic energy of the molecules, and d the path length in $\mathbf{B}$,

$$E = \tfrac{1}{2}mv^2 = m_p\frac{d^2}{t^2}, \tag{15}$$

since m, the mass of the molecule is twice the mass m_p of the proton. Thus

$$\mu_p = \frac{\pi\hbar}{B_0 d}\left(\frac{E}{m_p}\right)^{1/2} \tag{16}$$

$$= 1.424 \times 10^{-26}\ \mathrm{J\,T^{-1}} = 2.82\ \mu_N. \tag{17}$$

Comments

(1) It is straightforward to show that the state function $\psi(t)$ is an eigenfunction, with eigenvalue $M_S = 1$, of S_ρ, the operator for the component of spin in the direction $(\cos\rho, -\sin\rho, 0)$, where

$$\rho = \frac{\mu}{\hbar}Bt. \tag{18}$$

In other words, just as in the previous problem, the quantum mechanical calculation agrees with the classical result that the two protons may be regarded as a gyroscope, which, subjected to a couple by the magnetic field, precesses about the field with the Larmor frequency

$$\omega_L = \gamma B, \tag{19}$$

where the gyromagnetic ratio $\gamma = \mu/\hbar$. The precession is in the opposite sense to that of the neutron, because, for the proton, the magnetic moment and angular momentum vectors are in the same direction.

(2) Notice that, when the spin has precessed through an angle π, the molecule is in the state $M_S = -1$ (specified relative to the x axis). We might have guessed this from physical intuition. However, we might also have guessed that, when the spin had precessed through $\pi/2$, the molecule would be in the state $M_S = 0$. The quantum mechanical calculation shows this is not so; the state is actually a mixture of 50% $M_S = 0$, and 25% each of $M_S = \pm 1$. The expectation value $\langle S_x \rangle$, for a $\pi/2$ rotation, is of course zero.

8.3 (a) The state function at $t = 0$ is

$$\psi(0) = \frac{1}{\sqrt{(2s)}}\sum_n |n\rangle. \tag{1}$$

Therefore

$$\psi(t) = \frac{1}{\sqrt{(2s)}}\sum_n |n\rangle \exp(-\mathrm{i}E_n t/\hbar)$$

$$= \frac{1}{\sqrt{(2s)}}\sum_n |n\rangle \exp\{-\mathrm{i}(n + \tfrac{1}{2})\omega t\}, \tag{2}$$

since

$$E_n = (n + \tfrac{1}{2})\hbar\omega. \tag{3}$$

From (4.17) the displacement is

$$x = g(a + a^\dagger), \tag{4}$$

where a and $a^\dagger$ are the annihilation and creation operators for the harmonic oscillator, and

$$g = \left(\frac{\hbar}{2m\omega}\right)^{1/2}. \tag{5}$$

The expectation value of the displacement is

$$\langle x \rangle = g\langle \psi(t)|a + a^\dagger|\psi(t)\rangle \tag{6}$$

$$= \frac{g}{2s}\sum_{nn'}\langle n'|a + a^\dagger|n\rangle \exp\{\mathrm{i}(n' - n)\omega t\}. \tag{7}$$

The relations

$$a|n\rangle = \sqrt{n}|n-1\rangle, \quad a^\dagger|n\rangle = \sqrt{(n+1)}|n+1\rangle \tag{8}$$

show that the matrix element is zero unless $n' = n \pm 1$. Therefore

$$\langle x \rangle = \frac{g}{2s}\sum_n \sqrt{n}\exp(-\mathrm{i}\omega t) + \sqrt{(n+1)}\exp(\mathrm{i}\omega t). \tag{9}$$

The summation over n is from $N - s$ to $N + s$, where $N \gg s \gg 1$. So

$$\sqrt{n} \approx \sqrt{(n+1)} \approx \sqrt{N}. \tag{10}$$

There are $2s + 1 \approx 2s$ (approximately) equal terms in the summation; thus

$$\sum_n \sqrt{n}\exp(-\mathrm{i}\omega t) + \sqrt{(n+1)}\exp(\mathrm{i}\omega t) \approx 4s\sqrt{N}\cos\omega t, \tag{11}$$

giving

$$\langle x \rangle = \left(\frac{2\hbar N}{m\omega}\right)^{1/2}\cos\omega t. \tag{12}$$

(b) The equation of motion of a classical harmonic oscillator is

$$m\frac{\mathrm{d}^2x}{\mathrm{d}t^2} + kx = 0, \tag{13}$$

where the force constant $k = m\omega^2$. A solution of the equation is

$$x = X\cos\omega t, \tag{14}$$

where X is the amplitude of the oscillation. The total energy is

$$E = \tfrac{1}{2}kx^2 + \tfrac{1}{2}m\dot{x}^2 = \tfrac{1}{2}m\omega^2 X^2. \tag{15}$$

In the quantum calculation the energy is

$$E \approx (N + \tfrac{1}{2})\hbar\omega \approx N\hbar\omega. \tag{16}$$

Therefore, the quantum mechanical amplitude

$$\left(\frac{2\hbar N}{m\omega}\right)^{1/2} \approx \left(\frac{2E}{m\omega^2}\right)^{1/2}, \tag{17}$$

which, from (15), is equal to the classical amplitude X. So the expression for $\langle x \rangle$ in (12) is the same as the classical result in (14). We have thus shown that a coherent combination of a large number of energy eigenstates, with a spread of energies small compared to their mean energy ($1 \ll s \ll N$), behaves as a classical oscillator.

8.4 (a) First-order time-dependent perturbation theory gives the result that if a Hamiltonian $H^{(1)}$ is applied at $t = 0$ to a system in an initial state $|k\rangle$ with energy E_k, the probability that a transition has occurred to a state $|j\rangle$, with energy E_j at time t is $|c_j|^2$, where

$$c_j = \frac{1}{i\hbar}\int_0^t \langle j|H^{(1)}(t')|k\rangle \exp(i\omega t')\,dt', \quad \hbar\omega = E_j - E_k. \tag{1}$$

In the present problem $H^{(1)}(t')$ is the potential of the electron in the applied electric field, whose direction we take as the z axis, i.e.

$$H^{(1)}(t') = e\mathscr{E}_0 z \exp(-t'/\tau), \quad t' > 0. \tag{2}$$

For $t = \infty$, (1) becomes

$$c_j = \frac{e\mathscr{E}_0}{i\hbar}\langle j|z|k\rangle \int_0^\infty \exp\left\{\left(i\omega - \frac{1}{\tau}\right)t'\right\} dt'. \tag{3}$$

The time integral is

$$\int_0^\infty \exp\left\{\left(i\omega - \frac{1}{\tau}\right)t'\right\} dt' = \frac{1}{i\omega - \frac{1}{\tau}}\left[\exp\left\{\left(i\omega - \frac{1}{\tau}\right)t'\right\}\right]_0^\infty$$

$$= -\frac{1}{i\omega - \frac{1}{\tau}}. \tag{4}$$

(The factor $\exp(-t'/\tau)$ gives zero at the upper limit.) Therefore

$$c_j = \frac{e\mathscr{E}_0}{\hbar(\omega + \mathrm{i}/\tau)} \langle j|z|k \rangle. \tag{5}$$

The initial state is the ground state of the hydrogen atom, i.e. the quantum numbers are $n = 1$, $l = 0$, $m = 0$, and the final state has $n = 2$, $l = 1$, $m = 0$. We evaluate the matrix element in spherical polar coordinates r, θ, ϕ, using the relation $z = r\cos\theta$. Then

$$\langle j|z|k \rangle = \int_{\substack{\text{all}\\\text{space}}} u_{210}^* r \cos\theta \, u_{100} \mathrm{d(vol)}, \tag{6}$$

where

$$\mathrm{d(vol)} = r^2 \, \mathrm{d}r \sin\theta \, \mathrm{d}\theta \, \mathrm{d}\phi. \tag{7}$$

From Tables 4.1 and 4.2, pp. 52 and 54, the normalised wave functions for the initial and final states are

$$u_{100} = R_{10} Y_{00} = \frac{2}{\sqrt{(4\pi)}} \frac{1}{a_0^{3/2}} \exp(-r/a_0), \tag{8}$$

$$u_{210} = \mathrm{R}_{21} \mathrm{Y}_{10} = \frac{1}{\sqrt{(4\pi)}} \frac{1}{(2a_0)^{3/2}} \frac{r}{a_0} \exp(-r/2a_0) \cos\theta. \tag{9}$$

Thus

$$\langle j|z|k \rangle = \frac{1}{2^{3/2}} \frac{1}{a_0^4} \int_0^\infty r^4 \exp\left(-\frac{3r}{2a_0}\right) \mathrm{d}r \int_0^\pi \cos^2\theta \sin\theta \, \mathrm{d}\theta \tag{10}$$

$$= \frac{4!}{2^{3/2}} \left(\frac{2}{3}\right)^6 a_0. \tag{11}$$

(We have used the result $\int_0^\infty r^4 \exp(-\beta r) \, \mathrm{d}r = 4!/\beta^5$). From (5) and (11), the probability of a transition is

$$|c_j|^2 = \frac{2^{15}}{3^{10}} \frac{a_0^2 e^2 \mathscr{E}_0^2}{\hbar^2(\omega^2 + 1/\tau^2)}. \tag{12}$$

(b) The probability of a transition to the state 200 is zero. The parities of the functions u_{100} and u_{200} are positive, while that of z is negative. Therefore the parity of the product $u_{100} z u_{200}$ is negative; hence the matrix element, which is obtained by integrating $u_{100} z u_{200}$ over all space, is zero.

8.5 Suppose that, at $t = 0$, N_0 atoms are in an excited state k with energy E_k. At a later time t there are N atoms in the state k. From the definition of γ

$$dN = -\gamma N\, dt, \tag{1}$$

with solution

$$N = N_0 \exp(-\gamma t), \tag{2}$$

i.e. the probability of finding the atom in the excited state is

$$p = \exp(-\gamma t). \tag{3}$$

If there were no decay from the excited state, the state function would be

$$\psi(\mathbf{r}, t) = u_k(\mathbf{r}) \exp(-i\omega_k t), \tag{4}$$

where

$$\omega_k = E_k/\hbar. \tag{5}$$

We take account of the decay by writing (4) as

$$\psi(\mathbf{r}, t) = c_k(t) u_k(\mathbf{r}) \exp(-i\omega_k t), \tag{6}$$

where the probability of finding the atom in the state k is

$$|c_k(t)|^2 = \exp(-\gamma t), \tag{7}$$

i.e.

$$c_k(t) = \exp(-\gamma t/2) \qquad t > 0. \tag{8}$$

Therefore

$$\psi(\mathbf{r}, t) = u_k(\mathbf{r}) f(t), \tag{9}$$

where

$$f(t) = \exp\{(-\tfrac{1}{2}\gamma - i\omega_k)t\}. \tag{10}$$

The angular frequency (and hence energy) of the state k is no longer infinitely sharp, but has a spread. This may be calculated by expressing $f(t)$ as

$$f(t) = \int_{-\infty}^{\infty} g(\omega) \exp(-i\omega t)\, d\omega. \tag{11}$$

$g(\omega)$ is the Fourier transform of $f(t)$. The function $|g(\omega)|^2$ represents the 'amount' or 'strength' of the oscillation of angular frequency ω present in $f(t)$. From the theory of Fourier transforms

$$g(\omega) \propto \int_{-\infty}^{\infty} f(t) \exp(i\omega t)\, dt. \tag{12}$$

$f(t)$ is zero for $t < 0$, because the atom is excited at $t = 0$. So

$$g(\omega) \propto \int_0^\infty \exp\{(-\tfrac{1}{2}\gamma - i\omega_k + i\omega)t\}\,dt \tag{13}$$

$$= \frac{1}{\tfrac{1}{2}\gamma - i(\omega - \omega_k)}. \tag{14}$$

Thus

$$|g(\omega)|^2 \propto \frac{1}{(\omega - \omega_k)^2 + \tfrac{1}{4}\gamma^2}. \tag{15}$$

A function of this form is known as a *Lorentzian*. It has a maximum when $\omega = \omega_k$, and tends to zero as $|\omega - \omega_k|$ becomes large compared to γ. It is qualitatively similar to a Gaussian, but drops to zero less rapidly. It is plotted in Fig. 8.5.

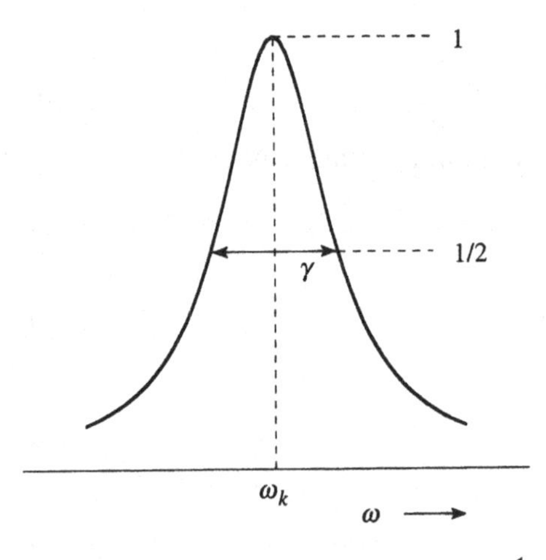

Fig. 8.5. *The Lorentzian function* $\dfrac{1}{(\omega - \omega_k)^2 + \frac{1}{4}\gamma^2}$.

When $\omega = \omega_k$, the value of $|g(\omega)|^2$ is $4/\gamma^2$, if we ignore the constant of proportionality. It drops to half this value when

$$\omega - \omega_k = \pm\tfrac{1}{2}\gamma. \tag{16}$$

The spread of the function, defined as the difference in the two values of ω for the half-height value, is thus γ.

8.6 The probability of a transition k (at time zero) to j (at time t) is

$$p_{k\to j}(t) = |c_{k\to j}(t)|^2, \tag{1}$$

where, from (8.9), the first-order expression for $c_{k\to j}(t)$ is

$$c_{k\to j}(t) = \frac{1}{\mathrm{i}\hbar}\int_0^t \langle j|H^{(1)}(t')|k\rangle \exp(\mathrm{i}\omega_{jk}t')\,\mathrm{d}t'. \tag{2}$$

The coefficient $c_{j\to k}(t)$ for the reverse transition is given by the same expression with k and j interchanged, i.e.

$$c_{j\to k}(t) = \frac{1}{\mathrm{i}\hbar}\int_0^t \langle k|H^{(1)}(t')|j\rangle \exp(\mathrm{i}\omega_{kj}t')\,\mathrm{d}t'. \tag{3}$$

Now $H^{(1)}(t')$ is a Hermitian operator. Therefore

$$\langle k|H^{(1)}(t')|j\rangle = \langle j|H^{(1)}(t')|k\rangle^*. \tag{4}$$

Also, from (8.6),

$$\hbar\omega_{kj} = E_k - E_j = -\hbar\omega_{jk}. \tag{5}$$

Therefore the integral in (3) is the complex conjugate of the one in (2). Thus

$$c_{j\to k}(t) = -\{c_{k\to j}(t)\}^*, \tag{6}$$

giving

$$p_{j\to k}(t) = |c_{j\to k}(t)|^2 = p_{k\to j}(t). \tag{7}$$

Comment

Although we have proved this result using first-order perturbation theory, it is generally true, i.e. the probability of transitions between two states, due to an external stimulus represented by $H^{(1)}$, is the same for transitions in either direction. The result is known as the *principle of detailed balancing*.

8.7 Denote the temperature of the walls of the box by T, and the equilibrium numbers of atoms in states k and j by N_k and N_j. Put the energy density function $I(\omega_{kj}) = I$. From the definitions of the Einstein coefficients A and B, the probability per unit time of a spontaneous transition $k\to j$ is A, and the probability per unit time of a stimulated transition $k\to j$ is BI. So the total number of transitions $k\to j$ per unit time is

$$n_{k\to j} = (A + BI)N_k. \tag{1}$$

Only stimulated transitions can occur in the reverse direction, so the number of transitions $j\to k$ per unit time is

$$n_{j\to k} = BIN_j. \tag{2}$$

Note that, from the result of the previous problem, the coefficients B in (1) and (2) are the same. In equilibrium

$$n_{k\to j} = n_{j\to k}, \tag{3}$$

so

$$\frac{N_j}{N_k} = \frac{A}{BI} + 1. \tag{4}$$

But the atoms are in temperature equilibrium with the walls of the box, and the ratio N_j/N_k is also given by the Boltzmann distribution. Thus

$$\frac{N_j}{N_k} = \frac{\exp(-E_j/k_B T)}{\exp(-E_k/k_B T)} = \exp(\hbar\omega_{kj}/k_B T), \tag{5}$$

where k_B is the Boltzmann constant, and we have used the result $\hbar\omega_{kj} = E_k - E_j$. From (4) and (5)

$$\frac{A}{BI(\omega_{kj})} = \exp(\hbar\omega_{kj}/k_B T) - 1. \tag{6}$$

Finally, the expression for $I(\omega_{kj})$ given by the Planck distribution law is

$$I(\omega_{kj}) = \frac{\hbar\omega_{kj}^3}{\pi^2 c^3} \frac{1}{\exp(\hbar\omega_{kj}/k_B T) - 1}. \tag{7}$$

Thus

$$\frac{A}{B} = \frac{\hbar\omega_{kj}^3}{\pi^2 c^3}. \tag{8}$$

Comment

This result, originally derived by Einstein, is a very useful one for obtaining an expression for the coefficient A by a semiclassical method. In this treatment, the electromagnetic field acting on the atom is not quantised, but is treated classically. The Einstein coefficient B is calculated by first-order time-dependent perturbation theory. The result in (8) then gives the coefficient A. The full treatment, in which the electromagnetic field is quantised, gives, not only the effect of the field on the atoms (stimulated emission and absorption), which we have considered, but also the effect of the atoms on the field (spontaneous emission), which we have not. The treatment, known as *quantum electrodynamics*, is considerably more advanced than the semiclassical method and gives the same expressions for the coefficients A and B. So we may accept the semiclassical theory as a reasonable compromise. The book by Schiff (1968) has an account of both methods.

8.8 In the electric dipole approximation, the Einstein coefficient A, giving the probability per unit time of a spontaneous transition from a state k, with energy E_k, to a state j, with energy E_j, is

$$A = \frac{e^2}{3\pi\varepsilon_0}\frac{1}{\hbar c^3}\omega_{kj}^3|\langle j|\mathbf{r}|k\rangle|^2, \tag{1}$$

where $\omega_{kj} = E_k - E_j$, and $\langle j|\mathbf{r}|k\rangle$ is the matrix element for the position operator $\mathbf{r}$.

For a pair of states in the hydrogen atom, $\langle j|\mathbf{r}|k\rangle$ is of the order of a_0, the Bohr radius. The energy $\hbar\omega_{kj}$ is of the order of the positive value of the energy of the ground state, i.e.

$$\omega_{kj} \sim \frac{1}{\hbar}\frac{e^2}{4\pi\varepsilon_0}\frac{1}{a_0}. \tag{2}$$

So

$$\begin{aligned} A &\sim \frac{e^2}{4\pi\varepsilon_0}\frac{1}{\hbar c^3}\left(\frac{1}{\hbar}\frac{e^2}{4\pi\varepsilon_0}\frac{1}{a_0}\right)^2\omega_{kj}a_0^2 \\ &= \left(\frac{e^2}{4\pi\varepsilon_0}\frac{1}{\hbar c}\right)^3\omega_{kj} \\ &= \alpha^3\omega_{kj}. \end{aligned} \tag{3}$$

For a transition to the ground state, the spread in the frequency comes from the spread in the energy of the excited state. The γ in Problem 8.5 is equal to A for an electric dipole transition. So the angular frequency width for the transition is $\gamma = A$. The mean angular frequency is ω_{kj}. The fractional frequency width is thus

$$\frac{A}{\omega_{kj}} \sim \alpha^3. \tag{4}$$

Since $\alpha \approx 0.007$, the value of the fraction is of the order of 10^{-6}.

Comment

The angular frequency width A is known as the *natural width* of the spectral line for the transition. However, the frequency of the line is further broadened by two other effects. The motion of the atoms causes the frequency to vary due to the Doppler effect. Secondly, the atoms collide with each other, and the effect of the collisions is to shorten the lifetime of the excited state. At room temperature and atmospheric pressure, the broadening due to these two effects is usually much larger than the natural width. Doppler broadening may be reduced by working at low temperatures, and collision broadening by reducing the pressure.

8.9 The ratio of stimulated to spontaneous emission processes is

$$r = \frac{BI}{A}, \tag{1}$$

where A and B are the Einstein coefficients for spontaneous and stimulated emission, and I is the energy density per unit range of angular frequency of the radiation at the ω value of the lamp. From (8.7.8)

$$\frac{BI}{A} = \frac{\pi^2 c^3}{\hbar \omega^3} I. \tag{2}$$

The output flux is

$$\begin{aligned} W &= (\text{energy density}) \times c \\ &= I \Delta\omega\, c, \end{aligned} \tag{3}$$

where $\Delta\omega$ is the spread in angular frequency of the radiation. If $\Delta\lambda$ is the spread in wavelength,

$$\frac{\Delta\lambda}{\lambda} = \frac{\Delta\omega}{\omega}. \tag{4}$$

From these equations we obtain

$$r = \frac{\lambda^4 W}{16\pi^2 c^2 \hbar} \frac{\lambda}{\Delta\lambda}. \tag{5}$$

Inserting the values $\lambda = 2.54 \times 10^{-7}$ m, $\Delta\lambda/\lambda = 10^{-5}$, $W = 10^3$ W m^{-2}, gives

$$r = 2.8 \times 10^{-4}.$$

Comment

For a conventional light source, such as a mercury lamp, the value of r is always small compared to 1, i.e. the rate of stimulated emission is much less than that of spontaneous emission. However, for lasers, and particularly for pulsed lasers, this is not so, and stimulated emission processes dominate. See Loudon (1983) for a detailed treatment of the quantum theory of light.

8.10 (a) The time derivative of the wave function $\psi_1 = n_1^{1/2} \exp(i\theta_1)$ is

$$\frac{\partial \psi_1}{\partial t} = \frac{1}{2n_1^{1/2}} \exp(i\theta_1) \frac{\partial n_1}{\partial t} + i n_1^{1/2} \exp(i\theta_1) \frac{\partial \theta_1}{\partial t}. \tag{1}$$

Also

$$\frac{1}{\mathrm{i}\hbar}(E_1\psi_1 + F\psi_2) = -\mathrm{i}\omega_1 n_1^{1/2}\exp(\mathrm{i}\theta_1) - \mathrm{i}\Omega n_2^{1/2}\exp(\mathrm{i}\theta_2), \tag{2}$$

where $\omega_1 = E_1/\hbar$, and $\Omega = F/\hbar$. Equating the right-hand sides of (1) and (2), and multiplying by $n_1^{1/2}\exp(-\mathrm{i}\theta_1)$ gives

$$\frac{1}{2}\frac{\partial n_1}{\partial t} + \mathrm{i}n_1\frac{\partial \theta_1}{\partial t} = -\mathrm{i}\omega_1 n_1 - \mathrm{i}\Omega(n_1 n_2)^{1/2}\exp\{\mathrm{i}(\theta_2 - \theta_1)\}. \tag{3}$$

Equating the real terms in (3) and using the fact that $n_1 \approx n_2 \approx n$, we obtain the result

$$\frac{\partial n_1}{\partial t} \approx 2\Omega n \sin(\theta_2 - \theta_1). \tag{4}$$

(b) Equate the imaginary terms in (3), divide by n_1, and again use the result $n_1 \approx n_2 \approx n$. This gives

$$\frac{\partial \theta_1}{\partial t} \approx -\omega_1 - \Omega\cos(\theta_2 - \theta_1). \tag{5}$$

If we start with $\psi_2 = n_2^{1/2}\exp(\mathrm{i}\theta_2)$, and repeat the whole calculation we obtain the similar result

$$\frac{\partial \theta_2}{\partial t} \approx -\omega_2 - \Omega\cos(\theta_2 - \theta_1), \tag{6}$$

where $\omega_2 = E_2/\hbar$. From (5) and (6)

$$\frac{\partial}{\partial t}(\theta_2 - \theta_1) = \omega_1 - \omega_2 = \frac{E_1 - E_2}{\hbar} = \frac{2eV}{\hbar}. \tag{7}$$

The last equation follows because, when a voltage V is applied across the junction, the energies E_1 and E_2 of the electron pairs in the two superconductors differ by an amount $2eV$.

There is a current of electron pairs between the two superconductors proportional to $\partial n_1/\partial t$, which we have shown in part (a) of the problem is proportional to $\sin(\theta_2 - \theta_1)$. Eq. (7) shows that

$$\theta_2 - \theta_1 = \frac{2eV}{\hbar}t + \delta, \tag{8}$$

where δ is a constant. So the current of superconducting electron pairs varies with angular frequency $2eV/\hbar$, i.e. with frequency

$$\nu = \frac{2eV}{h}. \tag{9}$$

Comments

(1) The present problem, taken from Kittel (1986), is a simplified treatment of the two Josephson effects. There is a current of superconducting electron pairs from superconductor 2 to 1, proportional to $\partial n_1/\partial t$, which is proportional to $\sin(\theta_2 - \theta_1)$. If there is no potential difference across the junction, $E_1 = E_2$, and (7) shows that $\theta_2 - \theta_1$ is then constant. Thus there is a constant current of density

$$J = J_0 \sin(\theta_2 - \theta_1), \tag{10}$$

where J_0 is a constant for the junction. This is the *dc Josephson effect*. The value of $\theta_2 - \theta_1$ cannot be set directly. Instead, a current is passed through the junction from a current generator, and, provided the current density is less than J_0, the phase difference adjusts itself to satisfy (10); the voltage across the junction is zero. If J is increased above J_0, the voltage across the junction starts to rise.

(2) The oscillation in the superconducting electron pair current that occurs when there is a dc potential difference across the junction, is known as the *ac Josephson effect*. The relation between the frequency and voltage depends only on the ratio e/h. Since this ratio is known to a high degree of accuracy (the fractional error in the experimental value is about 3×10^{-7}), and frequency can be measured more accurately than any other quantity in physics, the ac Josephson effect provides the most accurate method of measuring a dc voltage.

(3) The present simple treatment cannot give the value of J_0. Eq. (4) indicates that J_0 is proportional to Ω, i.e. to the quantity F, which is a measure of the interaction of the two superconductors across the insulating layer. We might expect F, and hence J_0, would increase with T_c, the critical temperature of the superconductors, and decrease with the thickness of the insulating layer. Qualitatively this is found to be the case.

9
Scattering, reactions

Summary of theory

1 Scattering theory

(a) Definitions of cross-sections

A beam of particles of mass m travelling along the z direction with velocity v is scattered by a short-range potential $V(\mathbf{r})$ centred on the origin – Fig. 9.1. We use spherical polar coordinates r, θ, ϕ, taking z as the polar axis. The Schrödinger equation is used to calculate the probability of a particle being scattered into a small solid-angle $\mathrm{d}\Omega$ in the direction θ, ϕ. The probability is expressed in terms of a *differential scattering cross-section* $\mathrm{d}\sigma/\mathrm{d}\Omega$, defined below, together with two other useful cross-sections.

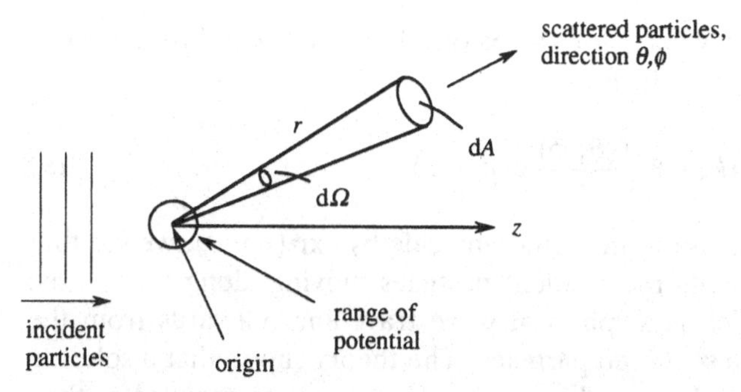

Fig. 9.1. *Diagram for relating the scattering cross-section* $\dfrac{\mathrm{d}\sigma}{\mathrm{d}\Omega}$ *to the wave function.*

Differential scattering cross-section

$$\frac{\mathrm{d}\sigma}{\mathrm{d}\Omega} = (\text{number of particles scattered per second into } \mathrm{d}\Omega \text{ in the direction } \theta,\ \phi)/F\mathrm{d}\Omega, \qquad (1)$$

where F is the flux of the incident particles.

Total scattering cross-section

$$\sigma_{sc} = (\text{total number of particles scattered per second})/F. \tag{2}$$

Absorption cross-section

$$\sigma_{abs} = (\text{number of particles absorbed per second})/F. \tag{3}$$

(b) Schrödinger equation

The particles have energy E, which is equal to their kinetic energy outside the range of potential. Thus

$$E = \tfrac{1}{2}mv^2 = \frac{\hbar^2 k^2}{2m}, \tag{4}$$

where

$$\hbar k = mv. \tag{5}$$

We have to solve the Schrödinger equation

$$\nabla^2\psi + \frac{2m}{\hbar^2}\{E - V(\mathbf{r})\}\psi = 0. \tag{6}$$

In the usual way ψ can be expressed as

$$\psi(\mathbf{r}, t) = u(\mathbf{r})\exp(-\mathrm{i}\omega t), \tag{7}$$

where

$$\hbar\omega = E. \tag{8}$$

We look for solutions where, for values of r large compared to the range of the potential, $u(\mathbf{r})$ has the form

$$u(\mathbf{r}) = \exp(\mathrm{i}kz) + \frac{f(\theta, \phi)}{r}\exp(\mathrm{i}kr). \tag{9}$$

If we multiply the terms on the right-hand side by $\exp(-\mathrm{i}\omega t)$, we see that the first term represents the incident particles moving along the z axis, and the second, which is a spherical wave travelling outwards from the origin, represents the scattered particles. The theory shows that a solution of the form of (9) can be found, provided $V(\mathbf{r})$ tends to zero faster than $1/r$ as r tends to infinity. We are confining the discussion to elastic scattering – hence the value of k is the same in the incident and scattered terms in (9). The function $f(\theta, \phi)$ is known as the *scattering amplitude*.

The differential scattering cross-section is related to $f(\theta, \phi)$ by

$$\frac{\mathrm{d}\sigma}{\mathrm{d}\Omega} = |f(\theta, \phi)|^2 \tag{10}$$

– see Problem 9.1.

(c) Born approximation

This is a first-order perturbation calculation, valid for fast particles and a weak potential, which means that the incident wave function is only slightly perturbed by the potential. The approximation gives

$$f(\theta, \phi) = -\frac{m}{2\pi\hbar^2}\int_{\substack{\text{all}\\ \text{space}}} V(\mathbf{r}) \exp(\mathrm{i}\boldsymbol{\kappa}\cdot\mathbf{r})\, \mathrm{d}\tau. \tag{11}$$

The quantity $\boldsymbol{\kappa}$, known as the *scattering vector*, is defined by $\boldsymbol{\kappa} = \mathbf{k} - \mathbf{k}'$, where $\mathbf{k}$ is the wavevector of the incident particles, and $\mathbf{k}'$ that of the scattered particles. The angle between $\mathbf{k}'$ and $\mathbf{k}$ is the scattering angle θ. Since the scattering is elastic,

$$|\mathbf{k}| = |\mathbf{k}'|, \quad \text{and} \quad \kappa = 2k \sin\frac{\theta}{2}. \tag{12}$$

(d) Method of partial waves

The method of partial waves involves a certain amount of mathematics, but the final results are relatively simple. The result is particularly simple for slow incident particles, when their wavelength is much larger than the range of the potential. We give here an outline of the reasoning, restricting the discussion to the case of a spherically symmetric potential $V(r)$.

The Schrödinger equation can be written as

$$\nabla^2 u + \left\{k^2 - \frac{2m}{\hbar^2}V(r)\right\}u = 0. \tag{13}$$

For a spherically symmetric potential the angular part of u is a spherical harmonic $Y_{lm}(\theta, \phi)$. From the symmetry of the problem, there is no dependence of the scattering on the azimuthal angle ϕ, which means that the spherical harmonic must have $m = 0$. The functions Y_{l0} are related to the Legendre polynomials $P_l(x)$, where $x = \cos\theta$, by

$$Y_{l0} = \left(\frac{2l+1}{4\pi}\right)^{1/2} P_l. \tag{14}$$

Putting

$$u(r, \theta) = \frac{\chi_l(r)}{r} P_l(\cos\theta), \tag{15}$$

and substituting in (13) leads to the result that $\chi_l(r)$ is a solution of the equation

$$\frac{\mathrm{d}^2\chi_l}{\mathrm{d}r^2} + \left\{k^2 - \frac{l(l+1)}{r^2} - \frac{2m}{\hbar^2}V\right\}\chi_l = 0. \tag{16}$$

The calculation proceeds in two stages. First the Schrödinger equation is solved for zero V. One solution is then $\exp(\mathrm{i}kz)$, which can be expressed as a linear combination of the terms in (15). We look for the asymptotic form, i.e. the linear combination that represents $\exp(\mathrm{i}kz)$ at large values of r. At large r, the term $l(l+1)/r^2$ in (16) tends to zero, and, for $V = 0$, the equation becomes

$$\frac{\mathrm{d}^2\chi_l}{\mathrm{d}r^2} + k^2\chi_l = 0, \tag{17}$$

which has solutions of the form $\sin(kr)$. The algebra shows that the linear combination of the sine functions that gives the asymptotic form of $\exp(\mathrm{i}kz)$ is

$$\exp(\mathrm{i}kz) = \sum_{l=0}^{l=\infty} A_l \frac{1}{r} \sin(kr - \tfrac{1}{2}l\pi) P_l(\cos\theta), \tag{18}$$

where

$$A_l = \frac{\mathrm{i}^l(2l+1)}{k}. \tag{19}$$

The significance of (18) can be seen if we express the sine function in its exponential form, and include the time-varying factor. The relevant term for a single l value is

$$\sin(kr - \tfrac{1}{2}l\pi)\exp(-\mathrm{i}\omega t) = \frac{1}{2\mathrm{i}}\exp\{\mathrm{i}(kr - \omega t - \tfrac{1}{2}l\pi)\}$$
$$- \frac{1}{2\mathrm{i}}\exp\{-\mathrm{i}(kr + \omega t - \tfrac{1}{2}l\pi)\}. \tag{20}$$

The terms on the right-hand side correspond to two spherical waves centred on the origin – the first an outgoing wave, and the second an incoming wave.

The second stage of the calculation is to make a similar expansion for the function $u(\mathbf{r})$, the solution of the Schrödinger equation when the potential is present. The equation satisfied by the radial function χ_l at large r is again (17), because the potential has a short range. So, the asymptotic form of χ_l is again $\sin(kr)$, and the asymptotic expansion of the function $u(\mathbf{r})$ is very similar to the right-hand side of (18). In fact the only difference is in the *phase* of the coefficients A_l. We can see this by rewriting (9) as

$$u(\mathbf{r}) - \exp(ikz) = \frac{f(\theta)}{r}\exp(ikr). \tag{21}$$

At large r the function $u(\mathbf{r})$, like $\exp(ikz)$, is a sum of outgoing and incoming waves, but the right-hand side of (21) corresponds to an outgoing wave only. Therefore the incoming waves for $u(\mathbf{r})$ must be identical with those of $\exp(ikz)$. Since the amplitudes of the outgoing and incoming waves are the same, the only effect of the scattering potential $V(r)$ on the wave function is to shift the phase of the outgoing asymptotic wave by an amount that we denote by δ_l.

The calculation shows that the function $f(\theta)$ is given by

$$f(\theta) = \frac{1}{2ik}\sum_{l=0}^{l=\infty}(2l+1)\{\exp(2i\delta_l) - 1\}P_l(\cos\theta). \tag{22}$$

Each l term is known as a *partial wave*. The same nomenclature is used as for atomic states. Thus the $l = 0$ wave is termed an s-wave, the $l = 1$ a p-wave, and so on. The lth wave, being an eigenfunction of the operator L^2, is a state in which the particle has orbital angular momentum about the origin of magnitude $\sqrt{\{l(l+1)\}}\hbar \approx l\hbar$. Since the linear momentum of the particle is $\hbar k$, a classical picture of the lth wave is that it corresponds to the particle passing the potential at a distance d_l from its centre, where $\hbar k\, d_l \approx l\hbar$, i.e.

$$d_l \approx \frac{l}{k}. \tag{23}$$

To calculate the phase shifts δ_l for a given potential $V(r)$ and energy E, we have to solve the Schrödinger equation in detail, which may involve considerable computation. (A simple example is given in Problem 9.6.) However, the classical picture leads to a useful qualitative result. Eq. (22) shows that, if $\delta_l = 0$, that partial wave gives no contribution to $f(\theta)$. If d_l, the distance of closest approach, is much larger than a, the range of the potential, the particle passes outside the potential and is not scattered, so $\delta_l = 0$ for that partial wave. The quantum calculation agrees with this. It follows from (23) that only those partial waves with $l \leqslant ak$ have non-zero values of δ_l, and contribute to the scattering.

A particularly simple case occurs if $ak \ll 1$, for then all the δ_l are zero, except δ_0. Since $k = 2\pi/\lambda$, where λ is the wavelength of the particle, the condition $ak \ll 1$ is equivalent to $\lambda \gg a$, i.e. the wavelength of the particle is large compared to the range of the potential. In this situation, when only the s-wave is disturbed, the angular dependence of $f(\theta)$ is given by the Legendre polynomial P_0, which is equal to 1, so the scattering is spherically symmetric. As k tends to zero, $\tan\delta_0$ becomes proportional to k, i.e. δ_0 tends to zero or a multiple of π, and the limiting value of $-\tan \delta_0/k$, denoted by b, is termed the *scattering length*. The concept is most commonly applied to the scattering of thermal neutrons.

2 Fermi golden rule

The Fermi golden rule is derived from first-order time-dependent perturbation theory. We use the same notation as in Chapter 8. The perturbing Hamiltonian $H^{(1)}$ brings about transitions between states j of the basic Hamiltonian $H^{(0)}$. Consider the case when $H^{(1)}$ is zero for time $t' < 0$, and constant in time thereafter. Up to time zero the system is in the state k with energy E_k. At time $t' = t$, the system has a finite probability of being in a state j with energy E_j, but the probability is negligible unless

$$|E_j - E_k| \leqslant \frac{\hbar}{t}. \tag{24}$$

If we add the probabilities for a transition to each state j, we obtain the probability P that the system makes a transition to *any* state j, and the theory shows that, provided the states are very closely spaced in energy, P is proportional to t. The Fermi golden rule is that the probability of a transition to any state j per unit time is then

$$\frac{\mathrm{d}P}{\mathrm{d}t} = \frac{2\pi}{\hbar}\rho(E_j)|\langle j|H^{(1)}|k\rangle|^2, \tag{25}$$

where $\rho(E_j)\mathrm{d}E_j$ is the number of j states in the energy range E_j to $E_j + \mathrm{d}E_j$. The calculation assumes that the energy range $\hbar/t$ is sufficiently small that, for the range of j satisfying (24), the density of states function $\rho(E_j)$, and the matrix element $\langle j|H^{(1)}|k\rangle$ are effectively constant.

Problems

The notation for the scattering problems in this chapter is given at the beginning of the summary of the theory, p. 179.

9.1 Particles travelling along the z axis are scattered by a short-range potential. If the wave function of the particles at large distances from the potential can be expressed in the form

$$u = \exp(\mathrm{i}kz) + \frac{1}{r} f(\theta, \phi) \exp(\mathrm{i}kr),$$

show that the differential scattering cross-section is

$$\frac{\mathrm{d}\sigma}{\mathrm{d}\Omega} = |f(\theta, \phi)|^2.$$

9.2 Show that, for a spherically symmetric potential,

$$\int_{\substack{\text{all}\\\text{space}}} V(\mathbf{r}) \exp(\mathrm{i}\boldsymbol{\kappa}\cdot\mathbf{r})\, \mathrm{d}\tau = \frac{4\pi}{\kappa} \int_0^\infty V(r) r \sin \kappa r \,\mathrm{d}r.$$

9.3 (a) Particles are incident on a spherically symmetric potential

$$V(r) = \frac{\beta}{r} \exp(-\gamma r),$$

where β and γ are constants. Show that, in the Born approximation, the differential scattering cross-section for the scattering vector $\boldsymbol{\kappa}$ is given by

$$\frac{\mathrm{d}\sigma}{\mathrm{d}\Omega} = \left\{\frac{2m\beta}{\hbar^2(\kappa^2 + \gamma^2)}\right\}^2.$$

(b) Use this result to derive the Rutherford formula for the scattering of α-particles, namely, that for α-particles of energy E incident on nuclei of atomic number Z, the differential scattering cross-section for scattering at an angle θ to the incident direction is

$$\frac{\mathrm{d}\sigma}{\mathrm{d}\Omega} = \left\{\frac{Ze^2}{8\pi\varepsilon_0 E \sin^2(\theta/2)}\right\}^2.$$

9.4 (a) A narrow beam of α-particles is incident normally on a thin gold foil, and the particles scattered at an angle θ to the initial direction are detected by a scintillator S that subtends a small solid angle $\mathrm{d}\Omega$ at C, the centre of the region where the α-particles strike the foil. Show that the fraction f of incident particles scattered into S is

$$f = \frac{d\sigma}{d\Omega} Nt \, d\Omega,$$

where $d\sigma/d\Omega$ is the differential scattering cross-section per nucleus for the angle θ, N is the number of nuclei per unit volume, and t, the thickness of the foil, is sufficiently small for the flux of incident particles to remain close to its original value at all depths of the foil.

(b) Systematic measurements of α-particle scattering were first performed by Geiger and Marsden in 1913. Their source was RaC′ (^{214}Po), which emits α-particles with an energy of E = 7.68 MeV. The scintillator had an area of 1 mm^2 and was 10 mm from the point C. For $t = 2.1 \times 10^{-7}$ m, and $\theta = 45°$, they found $f = 3.7 \times 10^{-7}$. What is the atomic number of gold from these measurements?
[The atomic weight of gold is 197, and its density is 19 300 kg m^{-3}.]

9.5 For neutrons of energy 25 meV, the values of the total scattering cross-section σ_{sc} and the absorption cross-section σ_{abs} for atoms of nitrogen and oxygen are

	σ_{sc}	σ_{abs}
nitrogen	11.5	1.8
oxygen	4.2	0.0

The units are 10^{-28} m^2.

Estimate the attenuation of a beam of neutrons of this energy in 1 m of dry air at a temperature of 20 °C.
[Density of air at 20 °C = 1.20 kg m^{-3}.]

9.6 Neutrons of mass m and energy E are incident on a spherically symmetric, square-well, attractive potential of depth W and range a, representing the nuclear force between the neutron and a nucleus. If the velocity $v \ll \hbar/ma$, show that

(a) the scattering is spherically symmetric,

(b) the s-wave phase shift δ satisfies

$$j \tan(ka + \delta) = k \tan ja,$$

where

$$k^2 = \frac{2mE}{\hbar^2}, \quad j^2 = \frac{2m(W + E)}{\hbar^2},$$

(c) the scattering length is

$$b = a\left(1 - \frac{\tan y}{y}\right),$$

where

$$y = (2mW)^{1/2}a/\hbar.$$

(d) What is the total scattering cross-section as E tends to zero?

9.7 (a) If the value of y for the potential of the last problem lies in the range $(n - \frac{1}{2})\pi < y < (n + \frac{1}{2})\pi$, where n is an integer, show that, for a neutron in this potential, there are n bound states (i.e. states of negative energy) with zero orbital angular momentum.

(b) Hence show that, if the scattering length is positive, there is at least one bound state.

9.8 (a) Show that, for a spherically symmetric potential $V(\mathbf{r}) = g\delta(\mathbf{r})$, where $\delta(\mathbf{r})$ is a 3-dimensional δ-function, and g is a constant, the Born approximation gives a scattering amplitude

$$f(\theta) = -\frac{m}{2\pi\hbar^2}g.$$

(b) Derive an expression for g, if the potential is used to simulate the nuclear scattering of thermal neutrons with scattering length b.

(c) Thermal neutrons of wavelength λ are incident on a slab of material whose nuclei have identical scattering lengths b. Show that the slab behaves as a medium with refractive index

$$n = 1 - \frac{1}{2\pi}Nb\lambda^2,$$

where N is the number of nuclei per unit volume.

(d) Show that, if b is positive, the critical glancing angle for total external reflection of the neutrons at the surface of the medium is

$$\gamma_c = \lambda(Nb/\pi)^{1/2}.$$

[The 3-dimensional δ-function has the properties

$$\delta(\mathbf{r}) = 0 \quad \mathbf{r} \neq 0, \quad \delta(\mathbf{r}) = \infty \quad \mathbf{r} = 0, \quad \int_{\substack{\text{all}\\\text{space}}} \delta(\mathbf{r})\,\mathrm{d}\tau = 1.$$

In part (c), relate the refractive index to the mean kinetic energy of the neutrons inside and outside the medium, and consider the value of the potential in the medium, averaged over a volume containing many nuclei.]

9.9 (a) Verify that, outside the range of a short-range potential, the wave function

$$u(r, \theta) = \frac{1}{r}\left(1 + \frac{\mathrm{i}}{kr}\right)\exp(\mathrm{i}kr)\cos\theta$$

represents an outgoing p-wave.

(b) A beam of particles represented by the plane wave $\exp(\mathrm{i}kz)$ is scattered by an impenetrable sphere of radius a, where $ka \ll 1$. By considering only s and p components in the scattered wave, show that, to order $(ka)^2$, the differential scattering cross-section for scattering at an angle θ is

$$\frac{\mathrm{d}\sigma}{\mathrm{d}\Omega} = a^2\{1 - \tfrac{1}{3}(ka)^2 + 2(ka)^2\cos\theta\}.$$

[The value of $\cos^2\theta$ averaged over all directions is $\frac{1}{3}$.]

9.10 A thermal neutron with velocity $v/2$ is absorbed by a proton with velocity $-v/2$, giving rise to a deuteron and a photon of energy E. The reverse process, in which a photon with energy E causes a deuteron with equal and opposite momentum to disintegrate into a neutron and a proton with velocities $v/2$ and $-v/2$, also occurs. Denote the cross-sections for the two processes by σ_{abs} and σ_{dis}.

(a) A box of volume Y contains a number of neutron–proton pairs and a number of photon–deuteron pairs. The centre-of-mass of each neutron–proton pair is at rest in the box, and the relative velocity of each pair is v. Similarly, the deuteron and photon in each pair have equal and opposite momenta, and the energy of each photon is E. Show that

$$\sigma_{\text{abs}} = \frac{2\pi}{\hbar} Y \frac{\rho_{\text{B}}}{v} |\langle \text{B}|H^{(1)}|\text{A}\rangle|^2,$$

where A denotes the state for a neutron–proton pair, B the state for a deuteron–photon pair, ρ_{B} is the number of B states per unit energy range, and $H^{(1)}$ is the perturbing Hamiltonian that brings about the transition in either direction.

(b) Hence show that

$$\frac{\sigma_{\text{abs}}}{\sigma_{\text{dis}}} = 6\left(\frac{E}{mvc}\right)^2,$$

where m is the mass of the neutron, assumed equal to that of the proton, and c is the speed of light.

Solutions

9.1 Consider a small element of area dA, perpendicular to the radius **r**, where r is large compared to the range of the potential – Fig. 9.1. The area dA subtends a solid angle dΩ at the origin, where

$$\mathrm{d}\Omega = \frac{\mathrm{d}A}{r^2}. \tag{1}$$

The term $u_\mathrm{i} = \exp(\mathrm{i}kz)$ in the wave function represents the incident particles with density

$$G_\mathrm{i} = |u_\mathrm{i}|^2 = 1, \tag{2}$$

while the term

$$u_\mathrm{s} = \frac{1}{r} f(\theta, \phi) \exp(\mathrm{i}kr)$$

represents the scattered particles with density

$$G_\mathrm{s} = |u_\mathrm{s}|^2 = \frac{|f(\theta, \phi)|^2}{r^2}. \tag{3}$$

The number of scattered particles passing through dA in unit time

$$= (\text{density of particles at } \mathbf{r}) \times v \times \mathrm{d}A$$

$$= \frac{|f(\theta, \phi)|^2}{r^2} v\, \mathrm{d}A = |f(\theta, \phi)|^2 v\, \mathrm{d}\Omega, \tag{4}$$

where v is the velocity of the incident and scattered particles. The incident flux is

$$F = G_\mathrm{i} \times v = v. \tag{5}$$

Inserting (4) and (5) into the definition of dσ/dΩ gives

$$\frac{\mathrm{d}\sigma}{\mathrm{d}\Omega} = |f(\theta, \phi)|^2. \tag{6}$$

Comments

(1) The function u_i is not normalised, because $|u_\mathrm{i}|^2 = 1$, which when integrated over all space gives infinity. However, the function gives a finite flux for the incident particles, and a finite cross-section.

Even so, the function $\exp(\mathrm{i}kz)$ cannot be an exact representation of the incident beam, because it implies that the beam has infinite dimensions perpendicular to the direction of the beam. In practice the cross-sectional area of the beam is, of course, finite. To represent a practical beam we need a wave packet. This consists of a set of plane waves, with a spread of wavevectors **k**. Strictly speaking, we should calculate u_{sc} for each individual **k**, and add the results. However, provided the dimensions of the beam are very large compared to the de Broglie wavelength, a condition normally well satisfied, the spread in the **k** values is very small compared to their mean value, and the result of the full calculation is usually almost the same as the one we have done. So we normally do not consider anything more complicated.

(2) A related point is that when we measure or detect the scattered particles, the detecting device must be outside the path of the incident beam – Fig. 9.2. This separation of the incident and scattered beams enables us to identify the terms u_i and u_s in the wave function as representing the incident and scattered particles.

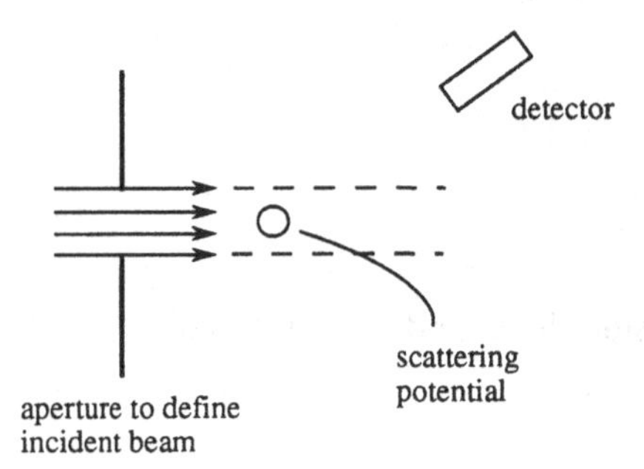

Fig. 9.2. *Geometry of scattering experiment.*

9.2 We are considering scattering at a fixed value of incident wavevector **k** and scattering angle θ. So the scattering vector $\boldsymbol{\kappa}$ is fixed, and we may take it as the polar axis in the integration over all space.

Since the potential is spherically symmetric, i.e. it depends only on the magnitude of **r**, we have

$$\int_{\substack{\text{all}\\\text{space}}} V(\mathbf{r}) \exp(\mathrm{i}\boldsymbol{\kappa}\cdot\mathbf{r})\,\mathrm{d}\tau = 2\pi\int_0^\infty V(r)r^2\,\mathrm{d}r\int_0^\pi \exp(\mathrm{i}\kappa r\cos\rho)\sin\rho\,\mathrm{d}\rho, \quad (1)$$

where ρ is the angle between **r** and the polar axis. Put $u = \cos\rho$. Then

$$\mathrm{d}u = -\sin\rho\,\mathrm{d}\rho, \quad (2)$$

and

$$\int_0^\pi \exp(\mathrm{i}\kappa r\cos\rho)\sin\rho\,\mathrm{d}\rho = \int_{-1}^1 \exp(\mathrm{i}\kappa r u)\,\mathrm{d}u = \frac{2}{\kappa r}\sin\kappa r. \quad (3)$$

Inserting (3) in (1) gives the result.

9.3 (a) The Born approximation gives the scattering amplitude

$$f(\theta) = -\frac{m}{2\pi\hbar^2} I, \tag{1}$$

where

$$I = \int_{\substack{\text{all}\\ \text{space}}} V(\mathbf{r}) \exp(\mathrm{i}\boldsymbol{\kappa}\cdot\mathbf{r})\, \mathrm{d}\tau. \tag{2}$$

Using the result of the last problem, with $V(r) = (\beta/r)\exp(-\gamma r)$, we obtain

$$I = \frac{4\pi\beta}{\kappa}\int_0^\infty \sin\kappa r \exp(-\gamma r)\, \mathrm{d}r. \tag{3}$$

The integral is readily evaluated by expressing the sine function as

$$\sin\kappa r = \frac{1}{2\mathrm{i}}\{\exp(\mathrm{i}\kappa r) - \exp(-\mathrm{i}\kappa r)\}, \tag{4}$$

and noting that the integrated expression vanishes at $r = \infty$ due to the factor $\exp(-\gamma r)$. The result is

$$I = \frac{4\pi\beta}{\kappa^2 + \gamma^2}. \tag{5}$$

Thus from (1) and (5)

$$\frac{\mathrm{d}\sigma}{\mathrm{d}\Omega} = |f(\theta)|^2 = \left\{\frac{2m\beta}{\hbar^2(\kappa^2 + \gamma^2)}\right\}^2. \tag{6}$$

(b) α-particles are scattered by the electrostatic interaction between the charge $2e$ on the α-particle and the charge Ze on the nucleus. The potential is

$$V(r) = \frac{2Ze^2}{4\pi\varepsilon_0}\frac{1}{r}. \tag{7}$$

This is represented by the potential in part (a), with

$$\beta = \frac{2Ze^2}{4\pi\varepsilon_0}, \quad \gamma = 0. \tag{8}$$

Inserting these values in (6) gives

$$\frac{\mathrm{d}\sigma}{\mathrm{d}\Omega} = \left\{\frac{mZe^2}{\pi\varepsilon_0\hbar^2\kappa^2}\right\}^2. \tag{9}$$

The energy of the α-particle is

$$E = \frac{p^2}{2m} = \frac{\hbar^2 k^2}{2m}. \tag{10}$$

where p is its momentum. Since $\kappa = 2k \sin(\theta/2)$,

$$\hbar^2 \kappa^2 = 8mE \sin^2 \frac{\theta}{2}. \tag{11}$$

Inserting (11) in (9) gives the required result.

Comment

Although we have obtained the Rutherford formula, the method is not a rigorous proof, because (9.9), the form of the solution to the Schrödinger equation on which the method is based, applies only to a potential $V(r)$ that tends to zero faster than $1/r$ as $r \to \infty$. This condition is satisfied by nuclear forces, which are short range, and also by the electrostatic force on a charged particle due to a nuclear charge screened by electrons, but it is not satisfied by a pure Coulomb potential. In this case it turns out that the function $\exp(ikz)$, which represents the incident particles in the absence of the potential, is not the correct asymptotic form for the incident particles when the potential is present. In other words, the Coulomb potential is so long-range that the incident wave is distorted even at infinity. The calculation for this potential is more complicated than that for the short-range potential; it is given in Chapter 3 of Mott and Massey (1965). The final result is the same as the one we have derived.

Rutherford obtained the same result in 1913 using classical mechanics. As we have noted before, a quantum mechanical result that does not contain the Planck constant can usually be derived by classical reasoning.

9.4 From (9.1) the number of particles scattered per second, through an angle θ, into a small solid-angle $d\Omega$, by a single nucleus, is $(d\sigma/d\Omega)F\,d\Omega$, where F is the incident flux of particles. If the particles strike a small area A of the foil of thickness t, there are NAt nuclei in the incident beam. Each nucleus scatters the α-particles independently. So the total number scattered per second into $d\Omega$ is

$$n_{sc} = \frac{d\sigma}{d\Omega} F\, d\Omega\, NAt. \tag{1}$$

The number of incident particles striking the foil per second is $n_{inc} = FA$. Therefore, the fraction of particles scattered into $d\Omega$ is

$$f = \frac{n_{sc}}{n_{inc}} = \frac{d\sigma}{d\Omega} Nt \, d\Omega. \tag{2}$$

Note that (1) assumes that the flux remains close to its initial value as the beam of particles traverses the foil. This is only true if the thickness of the foil is sufficiently small. Otherwise the flux decreases exponentially.

(b) Insert the values $E = 7.68$ MeV $= 7.68 \times 1.602 \times 10^{-13}$ J, and $\theta = 45°$ in the expression for $d\sigma/d\Omega$ in Problem 9.3b. This gives

$$\frac{d\sigma}{d\Omega} = 4.10 \times 10^{-31} \times Z^2 \, \text{m}^2. \tag{3}$$

The number of gold atoms per unit volume is

$$N = \frac{10^3 N_A \rho}{\text{AW}}, \tag{4}$$

where ρ is the density of gold, AW is its atomic weight, and N_A is the Avogadro constant. Putting $\rho = 19\,300$ kg m^{-3}, and AW = 197 gives

$$N = 5.90 \times 10^{28} \, \text{m}^{-3}. \tag{5}$$

The solid angle subtended by the scintillator at the point C is

$$d\Omega = \text{area of scintillator}/(\text{distance to } C)^2 = 10^{-2} \, \text{sr}. \tag{6}$$

Inserting (3), (5), and (6) in (2), together with $f = 3.7 \times 10^{-7}$, and $t = 2.1 \times 10^{-7}$ m, we obtain

$$Z = 85, \tag{7}$$

which may be compared with the correct value of 79.

Comment

With the exception of the value of E, the energy of the α-particles, the numerical values of the problem have been taken from the original experiments of Geiger and Marsden (1913), in which they measured the scattering of α-particles over a wide range of variables, and found it was in good agreement with the Rutherford formula. Geiger and Marsden overestimated the value of E by about 15%, which resulted in a similar overestimate for the value of Z. At the time of their measurements the value of the nuclear charge was not known, but they concluded, from results for seven different elements, that the Z value is about one half the atomic weight – a very good estimate for a pioneering experiment.

9.5 The total scattering cross-section σ_{sc} and the absorption cross-section σ_{abs} are defined on p. 180. A particle is removed from the beam if it is either scattered or absorbed. (We may neglect the very small fraction of particles that are scattered in the forward direction and remain in the beam.) If we put

$$\sigma_t = \sigma_{sc} + \sigma_{abs}, \tag{1}$$

then

$$\sigma_t = (\text{number of particles removed from the beam in unit time})/F, \tag{2}$$

where F is the flux of the incident particles.

Consider the neutron beam in the direction z, incident on a disc-shaped volume of air, of unit area and thickness dz – Fig. 9.3. Let F be the flux incident on the disc, and $F + dF$ the flux emerging from the disc. The neutrons are scattered by the nitrogen and oxygen nuclei in the air, and we may consider the two types of nuclei separately. The volume of the disc is dz, and it contains $n_N dz$ nitrogen atoms, where n_N is the number of nitrogen atoms per unit volume of air. From (2) the number of neutrons removed in unit time by the nitrogen nuclei in the disc is $\sigma_{tN} n_N F dz$, where σ_{tN} is the value of σ_t for nitrogen. A similar expression applies for oxygen. Thus the number of neutrons removed in unit time is

$$sF dz = -dF, \tag{3}$$

where

$$s = \sigma_{tN} n_N + \sigma_{tO} n_O. \tag{4}$$

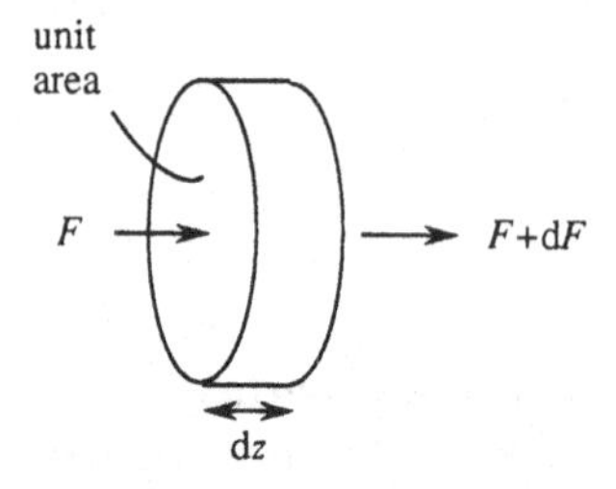

Fig. 9.3. *Attenuation of particle flux.*

Integrating (3) we have

$$F = F_0 \exp(-sz), \tag{5}$$

i.e. in traversing a distance z in air, the value of the flux drops by a factor $\exp(-sz)$.

Assume air is 80% nitrogen and 20% oxygen. Then $n_N = 4n_O$. The atomic weights of nitrogen and oxygen are 14 and 16. Thus the density of air is

$$\rho = \{(4 \times 14) + (1 \times 16)\}\frac{n_O}{10^3 N_A}, \tag{6}$$

where $N_A = 6.02 \times 10^{23}\,\text{mol}^{-1}$ is the Avogadro number. For $\rho = 1.20\,\text{kg}\,\text{m}^{-3}$, this gives

$$n_O = 1.003 \times 10^{25}\,\text{m}^{-3}. \tag{7}$$

For the values in the problem,

$$\sigma_{tN} = 13.3 \times 10^{-28}\,\text{m}^2, \quad \sigma_{tO} = 4.2 \times 10^{-28}\,\text{m}^2, \tag{8}$$

whence $s = 0.0576\,\text{m}^{-1}$, and, for $z = 1\,\text{m}$,

$$\exp(-sz) = 0.944, \tag{9}$$

i.e. the attenuation of the beam is 5.6% per metre.

Comments

(1) For many experiments with thermal neutrons, the attenuation of the beam by air is tolerable, so the measurements can be made without evacuating the path. This is of course not possible in experiments with charged particles, which produce such large ionisation that they must be done in a vacuum.

(2) Thermal neutrons are scattered and absorbed by nuclei. Therefore the cross-sections σ_{sc} and σ_{abs} depend on the particular isotopes. The values given in the problem are the mean values for the natural elements.

9.6 (a) The wavelength of the neutrons is $\lambda = h/mv$. So if

$$v = \frac{h}{m\lambda} \ll \frac{h}{ma}, \tag{1}$$

then $\lambda \gg a$, i.e. the wavelength of the particles is large compared to the range of the potential. For this condition only the s-partial wave is scattered, which means that the scattering is spherically symmetric.

(b) The wave $u(r)$ for the s-wave depends only on r, and may be written as

$$u(r) = \frac{\chi(r)}{r}, \tag{2}$$

where $\chi(r)$ satisfies the equation – see (9.16) –

$$\frac{d^2\chi}{dr^2} + \left\{k^2 - \frac{2m}{\hbar^2}V(r)\right\}\chi = 0. \tag{3}$$

The potential is shown in Fig. 9.4(a). Thus

$$\frac{d^2\chi}{dr^2} + j^2\chi = 0 \quad r < a, \qquad \frac{d^2\chi}{dr^2} + k^2\chi = 0 \quad r > a. \tag{4}$$

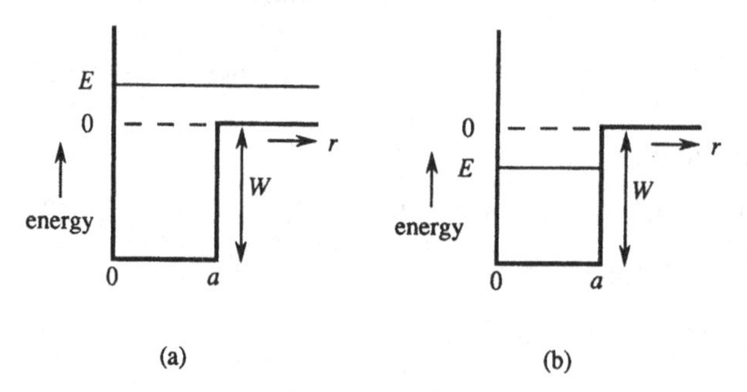

Fig. 9.4. *Radial square-well potential. (a) For $E > 0$, the particle is scattered – Problem 9.6. (b) For $E < 0$, the particle is bound by the potential – Problem 9.7.*

The required solution for the first equation is

$$\chi(r) = A \sin jr, \tag{5}$$

where A is a constant. It must be of this form (no cosine term), because the first equation applies at the origin, and to ensure that $u(r)$ is finite at the origin, $\chi(r)$ must be zero at that point. We write the solution of the second equation in the form

$$\chi(r) = B \sin(kr + \delta), \tag{6}$$

where B and δ are constants. Since the second equation does not apply at the origin, it is not necessary for its solution to be zero at that point. In the absence of the potential, the function $\chi(r)$ would have the form $B \sin kr$ for all values of r. So the δ in (6) is the change in phase of the asymptotic form of the s-wave due to the presence of the potential, i.e. it is the s-wave phase shift δ_0. The values of $\chi(r)$ and $\chi'(r)$ are continuous at $r = a$, which gives a relation for δ. Equating the values of the functions at $r = a$ in (5) and (6), we have

$$A \sin ja = B \sin(ka + \delta), \tag{7}$$

$$jA \cos ja = kB \cos(ka + \delta). \tag{8}$$

Dividing one by the other we obtain the required result

$$j \tan(ka + \delta) = k \tan ja. \tag{9}$$

(c) The scattering length b is defined by

$$b = -\lim_{k \to 0} \frac{\tan \delta}{k}. \tag{10}$$

From (9)

$$k \tan ja = j \frac{\tan ka + \tan \delta}{1 - \tan ka \tan \delta}, \tag{11}$$

whence

$$\tan \delta = \frac{k \tan ja - j \tan ka}{j + k \tan ka \tan ja}. \tag{12}$$

As $k \to 0$, $\tan ka \to ka$, and $ja \to (2mW)^{1/2} a/\hbar = y$, while the term $k \tan ka \tan ja$ tends to $k^2 a \tan ja$, which may be neglected in comparison to j. Thus

$$\tan \delta \to \frac{k \tan ja - kja}{j} \to ak \frac{\tan y - y}{y}. \tag{13}$$

From (10) and (13)

$$b = a\left(1 - \frac{\tan y}{y}\right). \tag{14}$$

(d) For s-wave scattering, (9.22) together with the result $P_0(\theta) = 1$, gives

$$f(\theta) = \frac{1}{2\mathrm{i}k}\{\exp(2\mathrm{i}\delta) - 1\} = \exp(\mathrm{i}\delta)\frac{\sin \delta}{k}. \tag{15}$$

The differential scattering cross-section is

$$\frac{\mathrm{d}\sigma}{\mathrm{d}\Omega} = |f(\theta)|^2 = \frac{\sin^2 \delta}{k^2}. \tag{16}$$

As $k \to 0$, $\tan \delta$ becomes proportional to k and thus tends to zero; δ tends to zero or a multiple of π. Then, from (10), (15), and (16),

$$f(\theta) \to -b, \quad \text{and} \quad \frac{\mathrm{d}\sigma}{\mathrm{d}\Omega} \to b^2. \tag{17}$$

The total scattering cross-section σ_{sc} is obtained by integrating $\mathrm{d}\sigma/\mathrm{d}\Omega$ over all directions in space. Since $\mathrm{d}\sigma/\mathrm{d}\Omega$ is independent of direction in this case, the result is simply

$$\sigma_{\mathrm{sc}} = 4\pi b^2 = 4\pi a^2 \left(1 - \frac{\tan y}{y}\right)^2. \tag{18}$$

Comment

The actual potential for the nuclear force between a nucleus and a neutron is not known in detail, and is not a static function of the separation distance. However, the force is certainly attractive and short-range, and the essential features of the nuclear scattering of thermal neutrons are correctly represented by the simple potential of the present problem. The range of the force a is of the order of a few times 10^{-15} m. The wavelength λ of thermal neutrons is of the order of 10^{-10} m. Since $k = 2\pi/\lambda$, ka is of the order of 10^{-4}. Except for rare cases where the quantity $y = (2mW)^{1/2}a/\hbar$ is very close to an odd multiple of $\pi/2$, the approximations made in the problem are valid, and the scattering amplitude $f(\theta)$ is a real constant, independent of θ and of k, i.e. of the velocity of the neutrons. The constant is $-b$, where b is the scattering length.

The scattering length of a nucleus depends on the particular nuclide, and also on the spin of the nucleus–neutron system. Our understanding of nuclear forces is not adequate to allow us to calculate the values of scattering lengths from known nuclear properties. Instead, we treat them as empirical quantities and determine their values experimentally. Unlike the corresponding quantity for the scattering of X-rays by an atom, which varies smoothly with atomic number Z (because X-rays are scattered by the *electrons* in the atom), nuclear scattering lengths vary erratically from one nucleus to another with neighbouring values of Z and N (neutron number). Most of the values are of the order of 10^{-14} m.

9.7 (a) For a neutron in a bound state of zero orbital angular momentum in the potential of the last problem, (9.6.2) and (9.6.3) apply as before. The only difference is that the energy E is now a negative quantity – see Fig. 9.4(b). The equations satisfied by the function $\chi(r)$ are

$$\chi'' + j^2\chi = 0 \quad r < a, \qquad \chi'' - \gamma^2\chi = 0 \quad r > a, \tag{1}$$

where

$$j^2 = \frac{2m}{\hbar^2}(W + E), \quad \gamma^2 = -\frac{2mE}{\hbar^2}. \tag{2}$$

Since E is negative, γ is real, and we take it positive. The two solutions are

$$\chi = A \sin jr, \quad \text{and} \quad \chi = B \exp(-\gamma r). \tag{3}$$

The function $\exp(\gamma r)$ is also a solution for $r > a$, but is rejected because it tends to infinity at large r. Equating $\chi(r)$ and $\chi'(r)$ at $r = a$, and dividing the two resulting equations, gives

$$\cot ja = -\frac{\gamma}{j}. \tag{4}$$

The values of E for a bound state are those for which j and γ satisfy this equation. At this point we may note that the equation is identical to (3.3.11). So the number n of bound states in the present problem is given by the number of odd-parity solutions in Problem 3.3. From the reasoning in part (c) of the problem, n is given by

$$(n - \tfrac{1}{2})\pi < y < (n + \tfrac{1}{2})\pi. \tag{5}$$

See Comment (1) below.

(b) Eq. (9.6.14) shows that the scattering length is positive if $\tan y < y$. This cannot be satisfied for $y < \pi/2$. So a positive scattering length means that $y > \pi/2$, and therefore, from (5) there is at least one bound state.

Comments

(1) Although the algebra for the odd-parity solutions of Problem 3.3 and the solutions of the present problem are the same, and lead to the same numerical conclusion, the physical situations are not the same. Problem 3.3 is a one-dimensional problem, and the function $u(x)$ in (3.3.1) is the actual wave function. Problem 9.6 is one in three dimensions, in which the orbital angular momentum is zero. This permits us to write the spherically symmetric wave function $u(r)$ as $\chi(r)/r$. The differential equation satisfied by $\chi(r)$ is the same as that satisfied by $u(x)$, and the boundary conditions for the former at $r = a$ are the same as those for the latter at $x = a$. However, at $r = 0$, we must have $\chi = 0$, so only the sine solution is permissible for $r < a$. In the one-dimensional problem, there is no such restriction on u. So both the cosine and sine solutions are permissible. For this reason, in one dimension, there is always at least one bound state for an attractive potential, no matter how weak it is, whereas, as we have seen, this is not true in three dimensions.

Although the algebra for the three-dimensional solution follows that of the odd-parity one-dimensional solution, the parity of the former is not odd. Since it is an s ($l = 0$) state, the parity is even.

(2) If the quantity y is equal to an odd number of $\pi/2$, then $\tan y$ is infinite and so is the value of the scattering length given by (9.6.14). This is a resonance potential, which gives rise to an infinite scattering cross-section. As y increases, the results of the present problem show that the number of bound states increases by unity every time y passes through one of the resonance values. The new bound state corresponds to

one of zero energy. It is easy to show that in this situation the range of the potential is exactly equal to an odd number of half-wavelengths of the sine wave solution.

(3) The scattering lengths of the proton provide a simple illustration of the ideas of the last two problems. Since the proton and the neutron both have spin $\frac{1}{2}$, the neutron–proton system has spin values 1 and 0. The former has degeneracy three, and is known as the triplet state; the latter is the singlet state. Each state has its own value of the scattering length. The measured values are $b_t = 10.8$ fm, and $b_s = -47.4$ fm. (1 fm = 10^{-15} m.)

Since b_t is positive, there must be at least one bound state for the neutron–proton system with spin 1. This corresponds to the deuteron. The latter is known to have no excited states, so the triplet system has exactly one bound state. Representing the interaction by our simple square-well potential, we see that the value of y lies between $\pi/2$ and π. If we assume a value for the range a, then the value of b_t, together with the relation (9.6.14), gives the value of y, and hence the depth of the well.

The singlet scattering length b_s is negative. This does not necessarily mean that there is no bound singlet state. But in fact such a state does not exist. So for the singlet potential the value of y is less than $\pi/2$. The fact that the numerical value of b_s is fairly large means that there is almost a bound state, i.e. if the singlet potential were slightly deeper or had a slightly longer range, a bound singlet state would exist.

9.8 (a) The scattering amplitude in the Born approximation is

$$f(\theta) = -\frac{m}{2\pi\hbar^2}I, \tag{1}$$

where

$$I = \int_{\substack{\text{all}\\ \text{space}}} V(\mathbf{r}) \exp(\mathrm{i}\boldsymbol{\kappa}\cdot\mathbf{r})\,\mathrm{d}\tau. \tag{2}$$

For $V(\mathbf{r}) = g\delta(\mathbf{r})$, the exponential term is evaluated at $\mathbf{r} = 0$. So $I = g$, and

$$f(\theta) = -\frac{m}{2\pi\hbar^2}g. \tag{3}$$

(b) Since g is a constant, $f(\theta)$ is a constant, which, for the nuclear scattering of thermal neutrons, is $-b$, where b is the scattering length. So, in the Born approximation, the delta function form of the potential gives the required scattering length b, provided we choose g to be

$$g = \frac{2\pi\hbar^2}{m_n} b, \tag{4}$$

where m_n is the mass of the neutron

(c) Neutrons travelling in a vacuum pass into a medium of scattering material as shown in Fig. 9.5(a). If the wavenumber of the neutrons is k_0 in the vacuum, and k in the medium, the refractive index of the medium is defined to be

$$n = \frac{k}{k_0}. \tag{5}$$

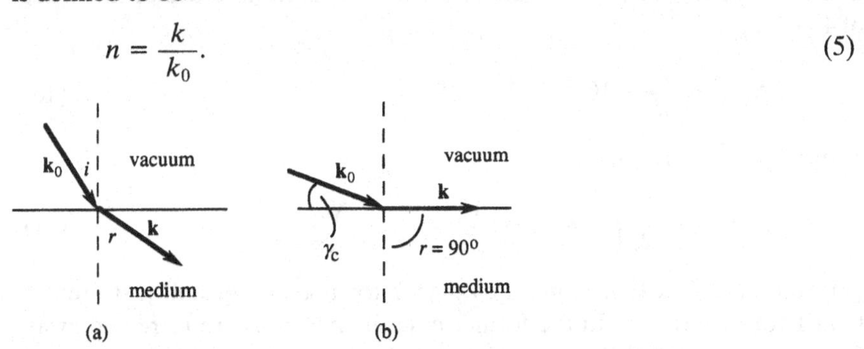

Fig. 9.5. *(a) Refraction of neutrons by medium with positive scattering length. (b) Critical external reflection.*

Each nucleus is at the centre of a potential which scatters the neutrons, and we take each potential to be of the δ-function form. The average value of the potential in the medium is obtained by integrating the potential function over a volume containing a large number of nuclei, and then dividing by the volume. Take the latter to be the unit volume. The integral of $g\delta(\mathbf{r})$ over all space is equal to g. So the average value of the potential in the medium is

$$\bar{V} = Ng = N\frac{2\pi\hbar^2}{m_n} b, \tag{6}$$

where N is the number of nuclei per unit volume.

Denote the energy of the neutrons in the vacuum by E. This energy is entirely kinetic. The kinetic energy of the neutrons in the medium is $E - \bar{V}$. Thus

$$k_0^2 = \frac{2m_n}{\hbar^2} E, \quad k^2 = \frac{2m_n}{\hbar^2}(E - \bar{V}). \tag{7}$$

From (5) and (7)

$$n = \left(1 - \frac{\bar{V}}{E}\right)^{1/2}. \tag{8}$$

From (6) and (7)

$$\frac{\bar{V}}{E} = \frac{4\pi Nb}{k_0^2} = \frac{Nb\lambda^2}{\pi}, \tag{9}$$

where $\lambda = 2\pi/k_0$ is the wavelength of the neutrons in vacuum. Now $N \sim 1/d^3$, where d, the average distance between the atoms in the scattering material, is of the order of 10^{-10} m. The wavelengths of thermal neutrons are also of this magnitude. Since the values of b are of the order of 10^{-14} m,

$$Nb\lambda^2 \sim \frac{b}{d} \sim 10^{-4}. \tag{10}$$

Therefore, $\bar{V} \ll E$, and

$$n = \left(1 - \frac{\bar{V}}{E}\right)^{1/2} \approx 1 - \frac{\bar{V}}{2E} = 1 - \frac{Nb\lambda^2}{2\pi}. \tag{11}$$

(d) Since $Nb\lambda^2 \ll 1$, n is just less than 1 for positive b, and just greater than 1 for negative b. In the former case the neutrons are refracted away from the normal at the surface of the medium as shown in Fig. 9.5(a). For critical external reflection we have – see Fig. 9.5(b) –

$$n = \frac{\sin i}{\sin r} = \frac{\cos \gamma_c}{1}. \tag{12}$$

Since n is close to 1, γ_c is small. Thus

$$\cos \gamma_c \approx 1 - \tfrac{1}{2}\gamma_c^2 \approx n = 1 - \frac{Nb\lambda^2}{2\pi}, \tag{13}$$

i.e.

$$\gamma_c = \left(\frac{Nb}{\pi}\right)^{1/2} \lambda. \tag{14}$$

Comments

(1) The condition for the validity of the Born approximation in scattering problems is that the wave function representing the incident particles should be only slightly disturbed by the scattering potential. This condition is not satisfied in the scattering of thermal neutrons, where the initial wave function is strongly perturbed. Nevertheless, provided the potential is represented by a δ-function, the approximation gives the correct result, namely, a scattering amplitude $f(\theta)$ which is a constant, independent of θ and of the neutron energy. This is the justification for its use, which greatly simplifies the theoretical treatment of thermal neutron scattering. Note that the δ-function potential is simply a

convenient mathematical device, and bears no relation to the form of the actual potential.

(2) The phenomenon of total external reflection of neutrons has important applications in experimental work. Measurement of the critical angle γ_c provides an accurate method for the experimental determination of the scattering length. Another application is in neutron guide tubes, which provide a means of guiding neutron beams over large distances by reflections from the inner surfaces of the tube. The principle is similar to that of guiding light beams with optic fibres.

9.9 (a) The dependence on θ for a p-wave ($l = 1$) is given by the Legendre polynomial $P_1(\theta) = \cos\theta$. So the angular part of the given function $u(r, \theta)$ has the required form for a p-wave. It remains to show that the radial part $R(r)$ has the correct form. Put $R(r) = \chi(r)/r$. From (9.16), outside the range of the potential, $\chi(r)$ satisfies

$$\frac{\mathrm{d}^2\chi}{\mathrm{d}r^2} + \left\{k^2 - \frac{l(l+1)}{r^2}\right\}\chi = 0, \tag{1}$$

with $l = 1$. For the given $u(r, \theta)$,

$$\chi(r) = \left(1 + \frac{\mathrm{i}}{kr}\right)\exp(\mathrm{i}kr). \tag{2}$$

Differentiating this twice with respect to r and substituting in (1) shows that the equation is satisfied with $l = 1$.

(b) Outside the range of the potential, the wave function for s and p waves has the form

$$u = \exp(\mathrm{i}kz) + \frac{A}{r}\exp(\mathrm{i}kr) + \frac{B}{r}\left(1 + \frac{\mathrm{i}}{kr}\right)\exp(\mathrm{i}kr)\cos\theta, \tag{3}$$

where A and B are constants. The term in A is the spherically symmetric s-wave, and the term in B is the p-wave. As $r \to \infty$

$$u \to \exp(\mathrm{i}kz) + \frac{f(\theta)}{r}\exp(\mathrm{i}kr). \tag{4}$$

Equating terms in $(1/r)\exp(\mathrm{i}kr)$ in (3) and (4) gives

$$f(\theta) = A + B\cos\theta. \tag{5}$$

Since the scattering object is an impenetrable sphere, the wave function vanishes on the surface $r = a$. Thus

$$\exp(\mathrm{i}ka\cos\theta) + \frac{A}{a}\exp(\mathrm{i}ka) + \frac{B}{a}\left(1 + \frac{\mathrm{i}}{ka}\right)\exp(\mathrm{i}ka)\cos\theta = 0. \tag{6}$$

The terms on the left-hand side that are independent of θ must sum to zero, and similarly the terms proportional to $\cos\theta$. Expanding $\exp(\mathrm{i}ka\cos\theta)$ in powers of ka we have

$$\exp(\mathrm{i}ka\cos\theta) = 1 + \mathrm{i}ka\cos\theta - \tfrac{1}{2}(ka)^2\cos^2\theta + 0(k^3a^3). \tag{7}$$

We cannot make the terms in $\cos^2\theta$ sum to zero in (6), because that would require the contribution from d-waves, which are excluded. However, we need to take into account the *average* value of $\cos^2\theta$, which is $\frac{1}{3}$. Summing the terms in (6) which are independent of θ gives

$$1 - \tfrac{1}{6}(ka)^2 + \frac{A}{a}\exp(\mathrm{i}ka) = 0, \tag{8}$$

whence

$$A = -a\{1 - \tfrac{1}{6}(ka)^2\}\exp(-\mathrm{i}ka). \tag{9}$$

Summing the terms in $\cos\theta$ gives

$$\mathrm{i}ka + \frac{B}{a}\left(1 + \frac{\mathrm{i}}{ka}\right)\exp(\mathrm{i}ka) = 0. \tag{10}$$

Since $ka \ll 1$, $\mathrm{i}/ka \gg 1$. Therefore to terms in $(ka)^2$

$$B = -a(ka)^2. \tag{11}$$

From (5), (9), and (11), the differential scattering cross-section is

$$\frac{\mathrm{d}\sigma}{\mathrm{d}\Omega} = |f(\theta)|^2 = |A + B\cos\theta|^2.$$

$$= a^2\{1 - \tfrac{1}{3}(ka)^2 + 2(ka)^2\cos\theta\}, \tag{12}$$

to terms in $(ka)^2$.

9.10 (a) Denote the number of neutron–proton pairs in the box by N_{A}, the number of deuteron–photon pairs by N_{B}, and the number of transitions $\mathrm{A} \to \mathrm{B}$ per unit time by $n_{\mathrm{A}\to\mathrm{B}}$. Then from the definition of a cross-section

$$\sigma_{\mathrm{abs}} = \sigma_{\mathrm{A}\to\mathrm{B}} = \frac{n_{\mathrm{A}\to\mathrm{B}}}{F_{\mathrm{A}}}, \tag{1}$$

where F_{A}, the neutron flux, is given by

$$F_{\mathrm{A}} = \frac{N_{\mathrm{A}}}{Y}v. \tag{2}$$

From the Fermi golden rule

$$n_{A\to B} = \frac{2\pi}{\hbar}\rho_B |\langle B|H^{(1)}|A\rangle|^2 N_A. \tag{3}$$

Therefore

$$\sigma_{A\to B} = \frac{2\pi}{\hbar} Y \frac{\rho_B}{v} |\langle B|H^{(1)}|A\rangle|^2. \tag{4}$$

(b) By the same reasoning we have

$$\sigma_{dis} = \sigma_{B\to A} = \frac{2\pi}{\hbar} Y \frac{\rho_A}{c} |\langle A|H^1|B\rangle|^2. \tag{5}$$

In deriving this equation from (4) we have replaced the relative velocity of the neutron and proton by the relative velocity of the deuteron and the photon, which is c, the speed of light. The perturbing Hamiltonian $H^{(1)}$ is the same for transitions in either direction. Since $H^{(1)}$ is a Hermitian operator

$$\langle A|H^{(1)}|B\rangle = \langle B|H^{(1)}|A\rangle^*, \tag{6}$$

i.e.

$$|\langle A|H^{(1)}|B\rangle|^2 = |\langle B|H^{(1)}|A\rangle|^2. \tag{7}$$

From (4), (5), and (7)

$$\frac{\sigma_{abs}}{\sigma_{dis}} = \frac{\rho_B}{\rho_A}\frac{c}{v}. \tag{8}$$

ρ_A is evaluated by the standard method of allowing only those neutron states for which the de Broglie waves are periodic in the box. (This is the same as the number of proton waves, because for each neutron–proton pair the momenta of the two particles are equal and opposite.) The result is

$$\rho_A = \frac{Y}{2\pi^2\hbar^3}\frac{p_a^2}{v} g_A. \tag{9}$$

In this expression, $p_a = \frac{1}{2}mv$ is the momentum of the neutron or proton in the centre-of-mass frame, and g_A, the spin degeneracy of each momentum state, is given by

$$g_A = (2I_n + 1)(2I_p + 1), \tag{10}$$

where I_n and I_p are the spins of the neutron and proton. Since $I_n = I_p = \frac{1}{2}$, $g_A = 4$. ρ_B is given by the expression analogous to (9), i.e.

$$\rho_B = \frac{Y}{2\pi^2\hbar^3}\frac{p_b^2}{c} g_B, \tag{11}$$

where p_b is the momentum of the deuteron or the photon. It is easier to calculate the latter. Since the energy of the photon in the centre-of-mass frame is E, its momentum is

$$p_b = \frac{E}{c}. \tag{12}$$

The spin of the deuteron is $I_d = 1$; thus its degeneracy is three. The analogue of spin states for the photon are its polarisation states, of which there are two. These may be regarded, either as states of linear polarisation along two perpendicular directions, or as left- and right-handed circular polarisations. So $g_B = 3 \times 2 = 6$.

Inserting (9) and (11) into (8), with the values of g_A and g_B, gives

$$\frac{\sigma_{abs}}{\sigma_{dis}} = 6\left(\frac{E}{mvc}\right)^2. \tag{13}$$

Comment

We normally do not measure E, the energy of the photon in the centre-of-mass frame, but rather E_{lab}, its energy in the laboratory frame in which the deuteron is at rest. It is straightforward to calculate the relation between E_{lab} and E. The result is

$$\frac{E_{lab}}{E} \approx 1 + \frac{E}{M_d c^2}, \tag{14}$$

where M_d is the mass of the deuteron. The energy of a thermal neutron is of the order of eV, so the energy of the photon in the reverse reaction is only just greater than the binding energy of the deuteron, which is about 2 MeV. The value of $M_d c^2$ is about 1860 MeV. Since the ratio of the cross-sections is proportional to E^2, the error in the ratio from assuming $E = E_{lab}$ is about 0.2%.

10 Miscellaneous

Problems

10.1 (a) By considering the quantity $I = \int (\lambda f + g)^*(\lambda f + g)\,\mathrm{d}\tau$, where f and g are general functions of position, and λ is a real constant, show that

$$\int f^* f\,\mathrm{d}\tau \int g^* g\,\mathrm{d}\tau \geqslant \tfrac{1}{4}\left\{\int (f^* g + g^* f)\,\mathrm{d}\tau\right\}^2.$$

This result is known as the *Schwarz inequality*.

(b) The operators A and B represent a pair of observables. When a system is in the state ψ, the observables have expectation values

$$\bar{a} = \int \psi^* A \psi\,\mathrm{d}\tau, \quad \bar{b} = \int \psi^* B \psi\,\mathrm{d}\tau,$$

and uncertainties Δa and Δb given by

$$(\Delta a)^2 = \int \psi^*(A - \bar{a})^2 \psi\,\mathrm{d}\tau, \quad (\Delta b)^2 = \int \psi^*(B - \bar{b})^2 \psi\,\mathrm{d}\tau.$$

Use the Schwarz inequality with

$$f = (A - \bar{a})\psi, \quad \text{and} \quad g = \mathrm{i}(B - \bar{b})\psi,$$

to show that

$$(\Delta a)^2(\Delta b)^2 \geqslant -\tfrac{1}{4}\left\{\int \psi^*(AB - BA)\psi\,\mathrm{d}\tau\right\}^2.$$

(c) Show that, for the observables position and linear momentum, this gives the uncertainty relation

$$\Delta p_x \Delta x \geqslant \frac{\hbar}{2}.$$

(d) Show that, if $\Delta p_x \Delta x = \hbar/2$, then ψ is a Gaussian.

10.2 The position probability function of a one-dimensional harmonic oscillator in equilibrium at temperature T is

$$f(x) = \frac{1}{Z}\sum_n \exp(-E_n/k_B T)u_n^2(x),$$

where u_n is the normalised eigenstate of the Hamiltonian for the quantum number n, E_n is the corresponding energy, and

$$Z = \sum_n \exp(-E_n/k_B T).$$

The quantity $f(x)\,dx$ is the probability of finding the particle with displacement between x and $x + dx$. The particle is to be regarded as a member of an ensemble of similar particles at temperature T, so it is not in a single state n, but in an incoherent mixture of states, each state being weighted by the appropriate Boltzmann factor. Show the following.

(a) $$\frac{df}{dx} = \frac{1}{Z}(2m\omega/\hbar)^{1/2}\sum_n \exp(-E_n/k_B T) \times \{\sqrt{n}u_{n-1}u_n - \sqrt{(n+1)}u_n u_{n+1}\},$$

$$xf = \frac{1}{Z}(\hbar/2m\omega)^{1/2}\sum_n \exp(-E_n/k_B T) \times \{\sqrt{n}u_{n-1}u_n + \sqrt{(n+1)}u_n u_{n+1}\},$$

where m is the mass, and ω the angular frequency.

(b) $f(x) = c\exp(-x^2/2\sigma^2)$,

where $\sigma^2 = (\hbar/2m\omega)\coth(\hbar\omega/2k_B T)$, and c is a constant.

(c) In the high-temperature limit ($k_B T \gg \hbar\omega$), the function $f(x)$ tends to the expected classical form.

10.3 A particle of mass m and charge e moves in the xy plane in a uniform magnetic field **B** along the z direction.

(a) Given that the Hamiltonian is

$$H = \frac{1}{2m}(\mathbf{p} - e\mathbf{A})^2,$$

where **p** is the linear momentum, and **A** is the vector potential, show that

$$H = \frac{1}{2m}\{p_x^2 + p_y^2 + eB(yp_x - xp_y) + \tfrac{1}{4}e^2B^2(x^2 + y^2)\}.$$

(b) Show that the operators b and $b^\dagger$ defined by

$$b = \frac{1}{\surd(2eB\hbar)}(\tfrac{1}{2}eBx + \mathrm{i}p_x + \tfrac{1}{2}\mathrm{i}eBy - p_y),$$

$$b^\dagger = \frac{1}{\surd(2eB\hbar)}(\tfrac{1}{2}eBx - \mathrm{i}p_x - \tfrac{1}{2}\mathrm{i}eBy - p_y),$$

satisfy the relations

$$bb^\dagger = \frac{H}{\hbar\omega} + \tfrac{1}{2}, \quad b^\dagger b = \frac{H}{\hbar\omega} - \tfrac{1}{2},$$

where $\omega = eB/m$.

(c) Hence show that the energy of the particle is

$$E = (n + \tfrac{1}{2})\hbar\omega,$$

where n is zero or a positive integer.

10.4 A one-dimensional harmonic oscillator has mass m and angular frequency ω. At time $t = 0$, a time-dependent state $\psi(x, t)$ of the oscillator is an eigenfunction of the annihilation operator a with eigenvalue μ (a complex number), i.e.

$$a\psi(x, 0) = \mu\psi(x, 0).$$

Show the following.

(a) $$\psi(x, 0) = c_0 \sum_{n=0}^{\infty} \frac{\mu^n}{\surd(n!)} u_n,$$

where u_n is the normalised eigenfunction of the Hamiltonian with quantum number n, and c_0 is a normalising constant.

(b) At a later time t, $\psi(x, t)$ is an eigenfunction of a with eigenvalue $\mu \exp(-\mathrm{i}\omega t)$.

(c) If $\mu = \lambda \exp(\mathrm{i}\rho)$, where λ and ρ are real, then

$$|\psi(x, t)|^2 = \frac{1}{\surd(2\pi)\sigma} \exp\{-(x - x_0)^2/2\sigma^2\},$$

where

$$x_0 = 2\sigma\lambda \cos(\rho - \omega t), \quad \sigma^2 = \frac{\hbar}{2m\omega}.$$

(d) Sketch the form of $|\psi(x, t)|^2$ at various times.

10.5 Consider a ring of N equally spaced ions, each with spin $\frac{1}{2}$, with an interaction between neighbouring ions given by the Hamiltonian

$$H = -\tfrac{1}{2}J\sum_{i=1}^{N}\boldsymbol{\sigma}_i\cdot\boldsymbol{\sigma}_{i+1},$$

where $\boldsymbol{\sigma}_i$ is the Pauli spin operator for the ith ion, and J is a constant. (An arbitrary ion is selected for $i = 1$.) A weak magnetic field is applied perpendicular to the ring, so all the σ_zs have the eigenfunctions α or β. The ground state χ_0 of the system corresponds to all the spins being in the state α. We denote by χ_j the state in which all the ions are in the state α, except the jth ion which is in the state β. Show the following.

(a) $$\boldsymbol{\sigma}_i\cdot\boldsymbol{\sigma}_{i+1}\chi_j = \chi_j \quad j \neq i \text{ or } i+1,$$

$$\boldsymbol{\sigma}_i\cdot\boldsymbol{\sigma}_{i+1}\chi_i = 2\chi_{i+1} - \chi_i,$$

$$\boldsymbol{\sigma}_i\cdot\boldsymbol{\sigma}_{i+1}\chi_{i+1} = 2\chi_i - \chi_{i+1}.$$

(b) If $\chi = \sum c_n\chi_n$ is an eigenfunction of the Hamiltonian with energy E, then

$$(E - E_0)c_n = J(2c_n - c_{n+1} - c_{n-1}),$$

where

$$E_0 = -\tfrac{1}{2}JN.$$

(c) If $c_n = (1/\sqrt{N})\exp(iqna)$, where a is the distance between the ions, then

$$E - E_0 = 2J(1 - \cos qa).$$

10.6 In a crystal lattice with orthorhombic symmetry, the electrostatic potential near a ${}^2P_{3/2}$ ion can be written as

$$V(\mathbf{r}) = Ax^2 + By^2 - (A + B)z^2,$$

where A and B are constants. A set of space wave functions for the free ion is

$$\psi_1 = R(r)Y_{11}, \quad \psi_0 = R(r)Y_{10}, \quad \psi_{-1} = R(r)Y_{1-1},$$

where $R(r)$ is a function that depends on the Hamiltonian of the ion, and Y_{11}, Y_{10}, Y_{1-1} are spherical harmonics.

(a) Show that the matrix of $V(\mathbf{r})$ on the basis of ψ_1, ψ_0, ψ_{-1} has the form

$$\gamma\begin{bmatrix} A+B & 0 & -A+B \\ 0 & -2(A+B) & 0 \\ -A+B & 0 & A+B \end{bmatrix},$$

where γ is a constant.

(b) Show that the matrix of $V(\mathbf{r})$ on the basis of the states $M_J = 3/2$, $1/2$, $-1/2$, $-3/2$ is

$$\begin{bmatrix} a & 0 & b & 0 \\ 0 & -a & 0 & b \\ b & 0 & -a & 0 \\ 0 & b & 0 & a \end{bmatrix},$$

where $a = \gamma(A + B)$, and $b = \gamma(B - A)/\sqrt{3}$.

(c) Assuming that the effect of $V(\mathbf{r})$ is small, show that the first-order changes in the energy of the ground state are $\pm\sqrt{(a^2 + b^2)}$, each value being doubly degenerate, and that one of the eigenstates for the positive energy change is $\cos(\theta/2)|3/2\rangle + \sin(\theta/2)|-1/2\rangle$, where $\theta = \tan^{-1}(b/a)$. What are the other eigenstates?
[Hint. In terms of Cartesian coordinates, the spherical harmonics for $l = 1$ are

$$Y_{11} = -(3/8\pi)^{1/2}(x + \mathrm{i}y)/r, \quad Y_{10} = (3/4\pi)^{1/2}z/r,$$
$$Y_{1-1} = (3/8\pi)^{1/2}(x - \mathrm{i}y)/r,$$

where

$$r^2 = x^2 + y^2 + z^2.]$$

10.7 A system of protons is subjected to a steady homogeneous magnetic field $\mathbf{B}$ along the z direction, and a rotating magnetic field in the xy plane with components $B_\mathrm{r}\cos\omega t$ along x, and $-B_\mathrm{r}\sin\omega t$ along y.

(a) Show that the Hamiltonian has the form

$$H = H^{(0)} + H^{(1)},$$

where

$$H^{(0)} = -\tfrac{1}{2}\hbar\omega_\mathrm{L}\sigma_z, \quad \omega_\mathrm{L} = \gamma_\mathrm{p}B,$$
$$H^{(1)} = -\tfrac{1}{2}\hbar\omega_\mathrm{r}(\sigma_x\cos\omega t - \sigma_y\sin\omega t), \quad \omega_\mathrm{r} = \gamma_\mathrm{p}B_\mathrm{r}.$$

σ_x, σ_y, σ_z are the Pauli spin operators for the proton, and γ_p is its gyromagnetic ratio.

(b) If the spin wave function is given by

$$\psi(t) = c_\alpha\exp(\mathrm{i}\omega_\mathrm{L}t/2)\alpha + c_\beta\exp(-\mathrm{i}\omega_\mathrm{L}t/2)\beta,$$

where α and β are the eigenfunctions of σ_z, show that the coefficients c_α and c_β satisfy the relations

$$\frac{dc_\alpha}{dt} = \tfrac{1}{2}i\omega_r \exp(iqt)c_\beta,$$

$$\frac{dc_\beta}{dt} = \tfrac{1}{2}i\omega_r \exp(-iqt)c_\alpha, \quad q = \omega - \omega_L.$$

(c) If at $t = 0$ all the protons are in the state α, show that the fraction in the state β at time t is given by

$$|c_\beta(t)|^2 = \frac{\omega_r^2}{q^2 + \omega_r^2} \sin^2\{\sqrt{(q^2 + \omega_r^2)}t/2\}.$$

(d) Compare this exact expression for $|c_\beta(t)|^2$ with that given by first-order perturbation theory.

10.8 A neutron interferometer is a device for splitting a monoenergetic beam of thermal neutrons into two spatially distinct beams and bringing them together again in a coherent combination – Fig. 10.1. While they are separate, a magnetic field may be applied to one of them so that, if the initial beam is polarised, the direction of the neutron spin may be rotated. Such an instrument is set up so that before recombination the two beams have equal intensities, and, in the absence of any magnetic fields, their effective path lengths are equal. Take x, y, z as a right-handed set of Cartesian axes, and denote the eigenstates of S_z, the operator corresponding to the z component of spin angular momentum, by α and β.

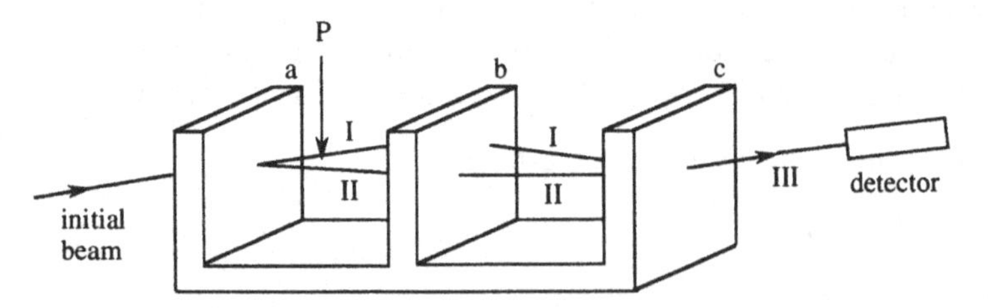

Fig. 10.1. *Neutron interferometer. The splitting and recombination of the neutron beams is done by Bragg reflection at the three plates,* a, b, c, *which are part of a single crystal of silicon. Bragg reflection has no effect on the spin, which remains in the same direction in space irrespective of a change in the direction of the neutron path. A magnetic field may be applied at* P *to rotate the spin in beam* I. *The two beams are combined coherently by plate* c *to give beam* III.

(a) Show that, if the incident beam is in the state α and a constant magnetic field **B**, along the x axis, is applied to one of the separated beams for a time $t = \pi\hbar/2\mu_n B$, where μ_n is the magnetic dipole moment of the neutron, the direction of the spin of the neutrons in this beam is reversed.

(b) The spin direction may also be reversed if **B** acts along the y axis for the same time. Is there any difference in the spin wave function of the neutrons after traversing the magnetic field in the two cases? What is the direction of the spin for the final combined beam in the two cases?

(c) If the magnetic field is applied along the x axis (i) for a time $t = \pi\hbar/\mu_n B$, and (ii) for a time $t = 2\pi\hbar/\mu_n B$, all other conditions being unchanged, what, in each case, would be the spin direction of the neutrons in the single beam after traversing the magnetic field? Would there be any difference in the intensity of the final combined beam in the two cases?

10.9 (a) Low-energy neutrons are scattered by a system of nuclei with spin J (not equal to zero). If $\hat{\mathbf{J}}$ and $\hat{\mathbf{S}}$ are the spin angular momentum operators (in units of $\hbar$) for the nucleus and neutron respectively, show that the eigenvalues of the operator $\hat{\mathbf{J}}.\hat{\mathbf{S}}$ are

$$\tfrac{1}{2}J \quad \text{for } T = J + \tfrac{1}{2}, \quad \text{and} \quad -\tfrac{1}{2}(J+1) \quad \text{for } T = J - \tfrac{1}{2},$$

where T is the spin of the nucleus–neutron system.

(b) Given that the operator for the neutron scattering length may be written in the form

$$\hat{b} = A + B\hat{\mathbf{J}}\cdot\hat{\mathbf{S}},$$

where A and B are constants, show that

$$A = \{(J+1)b^+ + Jb^-\}/(2J+1),$$

$$B = 2(b^+ - b^-)/(2J+1),$$

where b^+ and b^- are the scattering lengths (i.e. the eigenvalues of $\hat{b}$) for $T = J + \tfrac{1}{2}$ and $T = J - \tfrac{1}{2}$.

(c) Hence show that, for neutrons whose wavelengths are long compared with the distance between the protons of a hydrogen molecule, the total scattering cross-sections for ortho and parahydrogen are

$$\sigma_{\text{ortho}} = \frac{4\pi}{9}\{(3b^+ + b^-)^2 + 2(b^+ - b^-)^2\},$$

$$\sigma_{\text{para}} = \frac{4\pi}{9}(3b^+ + b^-)^2,$$

where b^+ and b^- are the triplet and singlet scattering lengths for the bound proton.
[The scattering length for a proton in a hydrogen molecule is $\tfrac{2}{3}$ times the bound scattering length; this is due to a reduced mass effect.]

10.10 A particle of mass m and charge e is constrained to move in a circular orbit of radius a. If x is the distance round the orbit, show that

(a) the functions

$$\frac{1}{\sqrt{(2\pi a)}}\exp(inx/a), \quad \frac{1}{\sqrt{(2\pi a)}}\exp(-inx/a),$$

where n is zero or a positive integer, are solutions of the Schrödinger equation with energy

$$E_0 = \frac{n^2\hbar^2}{2ma^2},$$

(b) the orbiting particle has magnetic dipole moment

$$\mu = \pm\frac{ne\hbar}{2m}.$$

A uniform magnetic field **B** is applied perpendicular to the orbit. Show that

(c) the field may be represented by a vector potential **A**, which acts tangentially round the orbit, and has magnitude $A = aB/2$,

(d) for a fixed value of n, the two functions in (a) remain solutions of the Schrödinger equation, with energies

$$E = E_0 + E_1 + E_2,$$

where

$$E_1 = \mp\frac{n\hbar^2 j}{ma}, \quad E_2 = \frac{\hbar^2 j^2}{2m}, \quad \hbar j = eA,$$

(e) the quantity E_1 corresponds to the paramagnetic energy, and E_2 to the diamagnetic energy.

Solutions

10.1 (a) Put

$$u = \int f^* f \,\mathrm{d}\tau, \quad v = \int f^* g \,\mathrm{d}\tau, \quad w = \int g^* g \,\mathrm{d}\tau. \tag{1}$$

Then

$$I = \int (\lambda f + g)^*(\lambda f + g)\,\mathrm{d}\tau = u\lambda^2 + (v + v^*)\lambda + w. \tag{2}$$

Since the integrand of I cannot be negative at any point, $I \geqslant 0$. The equality holds only if $\lambda f + g$ is everywhere zero. Therefore

$$u\lambda^2 + (v + v^*)\lambda + w \geqslant 0. \tag{3}$$

But λ is real. So the inequality (3) is equivalent to the condition that the quadratic equation $u\lambda^2 + (v + v^*)\lambda + w = 0$ has no real roots, or two equal roots, which is

$$4uw \geqslant (v + v^*)^2, \tag{4}$$

i.e.

$$\int f^* f \,\mathrm{d}\tau \int g^* g \,\mathrm{d}\tau \geqslant \tfrac{1}{4}\left\{\int (f^* g + g^* f)\,\mathrm{d}\tau\right\}^2. \tag{5}$$

(b) With f and g as defined in the problem, the left-hand side of the Schwarz inequality becomes

$$\int \{(A - \bar{a})\psi\}^*(A - \bar{a})\psi \,\mathrm{d}\tau \int \{\mathrm{i}(B - \bar{b})\psi\}^*\mathrm{i}(B - \bar{b})\psi \,\mathrm{d}\tau \tag{6}$$

$$= \int \psi^*(A - \bar{a})^2\psi \,\mathrm{d}\tau \int \psi^*(B - \bar{b})^2\psi \,\mathrm{d}\tau \tag{7}$$

$$= (\Delta a)^2(\Delta b)^2. \tag{8}$$

In the step from (6) to (7) we have used the fact that A and B are Hermitian, and therefore $A - \bar{a}$ and $B - \bar{b}$ are Hermitian.

The right-hand side of the inequality is

$$\tfrac{1}{4}\mathrm{i}^2\left[\int [\{(A - \bar{a})\psi\}^*(B - \bar{b})\psi - \{(B - \bar{b})\psi\}^*(A - \bar{a})\psi]\,\mathrm{d}\tau\right]^2 \tag{9}$$

$$= -\tfrac{1}{4}\int \psi^*\{(A - \bar{a})(B - \bar{b}) - (B - \bar{b})(A - \bar{a})\}\psi \,\mathrm{d}\tau \tag{10}$$

$$= -\tfrac{1}{4}\int \psi^*(AB - BA)\psi \,\mathrm{d}\tau. \tag{11}$$

The other terms in (10) sum to zero. In going from (9) to (10) we have again made use of the fact that $A - \bar{a}$ and $B - \bar{b}$ are Hermitian. Inserting (8) and (11) in the Schwarz inequality we have

$$(\Delta a)^2(\Delta b)^2 \geqslant -\tfrac{1}{4}\int \psi^*(AB - BA)\psi\,\mathrm{d}\tau. \tag{12}$$

(c) For the observables position and linear momentum

$$A = \frac{\hbar}{\mathrm{i}}\frac{\mathrm{d}}{\mathrm{d}x}, \qquad B = x, \tag{13}$$

$$AB - BA = \frac{\hbar}{\mathrm{i}}. \tag{14}$$

So

$$(\Delta p_x)^2(\Delta x)^2 \geqslant -\tfrac{1}{4}(\hbar/\mathrm{i})^2\int \psi^*\psi\,\mathrm{d}x = \tfrac{1}{4}\hbar^2, \tag{15}$$

since ψ is normalised. Thus

$$\Delta p_x \Delta x \geqslant \tfrac{1}{2}\hbar. \tag{16}$$

(d) Put $\bar{p}_x = \bar{x} = 0$, which does not affect the argument. Then

$$f = \frac{\hbar}{\mathrm{i}}\frac{\mathrm{d}\psi}{\mathrm{d}x}, \qquad g = \mathrm{i}x\psi. \tag{17}$$

The equality sign in (16) holds only if $\lambda f + g = 0$ for all values of x, where λ is a real constant. So the condition for equality is

$$\lambda\frac{\hbar}{\mathrm{i}}\frac{\mathrm{d}\psi}{\mathrm{d}x} = -\mathrm{i}x\psi, \quad \text{i.e.} \quad \frac{\mathrm{d}\psi}{\mathrm{d}x} = \frac{1}{\lambda\hbar}x\psi, \tag{18}$$

the solution of which is

$$\psi = c\exp(-x^2/4\Delta^2), \tag{19}$$

where c is a constant, and Δ^2 (a positive quantity) satisfies

$$\Delta^2 = -\frac{\lambda\hbar}{2}. \tag{20}$$

We could have written the solution without the minus signs in (19) and (20). However such a solution would tend to infinity at large x, and is therefore excluded. The parameter λ is real and negative.

Comment

In Problem 2.3 we showed that for a Gaussian wave function the equality sign holds in (16). We have now shown that if the equality sign holds, the wave function must be Gaussian. Since the quantity λ in (20) is real, so also is Δ^2. A Gaussian with a complex value for Δ^2 does not give uncertainties satisfying the equality condition. We have had an illustration of this in Problem 2.8. At time $t = 0$, the wave function is a Gaussian with a real value for Δ^2, but as it develops with time Δ^2 becomes complex. At $t = 0$, $\Delta = \Delta_0$, and the equality condition is satisfied, but with increasing time Δ^2 increases according to (2.8.8). The uncertainty in p_x remains constant as shown in part (b) of the problem. Thus the value of $\Delta p_x \Delta x$ increases.

10.2 (a) Use the annihilation and creation operators

$$a = (2m\hbar\omega)^{-1/2}(m\omega x + ip), \quad a^\dagger = (2m\hbar\omega)^{-1/2}(m\omega x - ip). \tag{1}$$

Replacing p by $(\hbar/\mathrm{i})\,\mathrm{d}/\mathrm{d}x$, and rearranging we obtain

$$x = (\hbar/2m\omega)^{1/2}(a + a^\dagger), \quad \frac{\mathrm{d}}{\mathrm{d}x} = (m\omega/2\hbar)^{1/2}(a - a^\dagger). \tag{2}$$

Also

$$au_n = \sqrt{n}\,u_{n-1}, \quad a^\dagger u_n = \sqrt{(n+1)}\,u_{n+1}. \tag{3}$$

Then

$$\frac{\mathrm{d}f}{\mathrm{d}x} = \frac{2}{Z}\sum_n \exp(-E_n/k_B T)u_n \frac{\mathrm{d}u_n}{\mathrm{d}x}$$

$$= \frac{1}{Z}(2m\omega/\hbar)^{1/2}\sum_n \exp(-E_n/k_B T)u_n(au_n - a^\dagger u_n)$$

$$= \frac{1}{Z}(2m\omega/\hbar)^{1/2}\sum_n \exp(-E_n/k_B T)$$

$$\times \{\sqrt{n}\,u_n u_{n-1} - \sqrt{(n+1)}\,u_n u_{n+1}\}. \tag{4}$$

Similarly

$$xf = \frac{1}{Z}(\hbar/2m\omega)^{1/2}\sum_n \exp(-E_n/k_B T)u_n(au_n + a^\dagger u_n)$$

$$= \frac{1}{Z}(\hbar/2m\omega)^{1/2}\sum_n \exp(-E_n/k_B T)$$

$$\times \{\sqrt{n}\,u_{n-1}u_n + \sqrt{(n+1)}\,u_n u_{n+1}\}. \tag{5}$$

(b) In the sum over n in (4), a specific $\sqrt{(n+1)}\,u_n u_{n+1}$ term occurs twice, once multiplied by $\exp(-E_n/k_B T)$, and once multiplied by

$$\exp(-E_{n+1}/k_B T) = \exp(-\hbar\omega/k_B T)\exp(-E_n/k_B T). \tag{6}$$

Therefore

$$\frac{\mathrm{d}f}{\mathrm{d}x} = -(2m\omega/\hbar)^{1/2}\{1 - \exp(-\hbar\omega/k_B T)\}S, \tag{7}$$

where

$$S = \frac{1}{Z}\sum_n \exp(-E_n/k_B T)\sqrt{(n+1)}\,u_n u_{n+1}. \tag{8}$$

Similarly

$$xf = (\hbar/2m\omega)^{1/2}\{1 + \exp(-\hbar\omega/k_B T)\}S. \tag{9}$$

From (7) and (9)

$$\frac{\mathrm{d}f}{\mathrm{d}x} = -\frac{1}{\sigma^2}xf, \tag{10}$$

where

$$\sigma^2 = \frac{\hbar}{2m\omega}\,\frac{1 + \exp(-\hbar\omega/k_B T)}{1 - \exp(-\hbar\omega/k_B T)} = \frac{\hbar}{2m\omega}\coth\frac{\hbar\omega}{2k_B T}. \tag{11}$$

The solution of (10) is

$$f(x) = c\exp(-x^2/2\sigma^2), \tag{12}$$

where c is a constant, which, if needed, is fixed by the normalisation condition

$$\int_{-\infty}^{\infty} f(x)\,\mathrm{d}x = 1. \tag{13}$$

σ is the standard deviation of $f(x)$, and we see from (11) that it increases with T, as expected.

(c) When $k_B T \gg \hbar\omega$,

$$\coth\frac{\hbar\omega}{2k_B T} \to \frac{2k_B T}{\hbar\omega}, \quad \text{and} \quad \sigma^2 \to \frac{k_B T}{m\omega^2}. \tag{14}$$

Thus

$$f(x) \to c\exp(-m\omega^2 x^2/2k_B T). \tag{15}$$

For a particle in a harmonic potential, the potential energy is

$$V(x) = \tfrac{1}{2}m\omega^2 x^2. \tag{16}$$

Thus

$$f(x) = c\exp\{-V(x)/k_B T\}. \tag{17}$$

Classically $f(x)$ is given by the Boltzmann factor, which is just the expression in (17).

Comments

(1) The result that the position probability function for a harmonic oscillator is a Gaussian was first derived by Bloch (1932). It has an important application in the diffraction of X-rays, neutrons, and electrons from crystals. If the atoms in the crystal were rigidly fixed at their equilibrium positions in the lattice, the waves representing the scattered particles would give constructive interference in discrete, extremely sharp, directions in space, specified by Bragg's law. However, owing to their thermal energy, the atoms are vibrating about their equilibrium positions. Constructive interference occurs over a small range of directions around the exact Bragg direction, and the scattering in that direction is correspondingly reduced. The term in the cross-section representing this reduction is known as the *Debye–Waller factor*. For a crystal with harmonic interatomic forces, the Debye–Waller factor is a Gaussian function of the scattering vector $\boldsymbol{\kappa}$, which may be shown to be a consequence of the Gaussian form of the probability distribution function for the atomic displacements.

(2) In part (c) of the problem we looked at the form of $f(x)$ at high temperature. We may also consider its form at low temperature, i.e. when $k_B T \ll \hbar\omega$. Then

$$\coth\frac{\hbar\omega}{2k_B T} \to 1, \quad \text{and} \quad \sigma^2 \to \frac{\hbar}{2m\omega}, \tag{18}$$

giving

$$f(x) = c\exp\left(-\frac{m\omega}{\hbar}x^2\right). \tag{19}$$

Comparing this with (4.9.23) we see that $f(x) = u_0^2$. This is to be expected. When $k_B T \ll \hbar\omega$, the probability of the particle being in any state other than the ground state is effectively zero.

10.3 (a) The magnetic field $\mathbf{B}$ is related to the vector potential by $\mathbf{B} = \operatorname{curl}\mathbf{A}$. So for a magnetic field in the z direction we require

$$(\operatorname{curl}\mathbf{A})_x = 0, \quad (\operatorname{curl}\mathbf{A})_y = 0, \quad (\operatorname{curl}\mathbf{A})_z = B. \tag{1}$$

The components of curl are given by

$$(\operatorname{curl} A)_x = \frac{\partial A_z}{\partial_y} - \frac{\partial A_y}{\partial z}, \tag{2}$$

with cyclic permutation of x, y, z for the other two. It can be seen that the equations in (1) are satisfied by

$$A_x = -\tfrac{1}{2}By, \quad A_y = \tfrac{1}{2}Bx, \quad A_z = 0. \tag{3}$$

The Hamiltonian is

$$\begin{aligned} H &= \frac{1}{2m}(\mathbf{p} - e\mathbf{A})^2 \\ &= \frac{1}{2m}(p^2 - 2e\mathbf{p}\cdot\mathbf{A} + e^2\mathrm{A}^2). \end{aligned} \tag{4}$$

Substitute the values of (3), together with $p_z = 0$, into (4). This gives

$$H = \frac{1}{2m}\{p_x^2 + p_y^2 + eB(yp_x - xp_y) + \tfrac{1}{4}e^2B^2(x^2 + y^2)\}. \tag{5}$$

(b) We have

$$\begin{aligned} bb^\dagger &= \frac{1}{2eB\hbar}(\tfrac{1}{2}eBx + \mathrm{i}p_x + \tfrac{1}{2}\mathrm{i}eBy - p_y) \\ &\qquad \times (\tfrac{1}{2}eBx - \mathrm{i}p_x - \tfrac{1}{2}\mathrm{i}eBy - p_y) \\ &= \frac{1}{2eB\hbar}\{\tfrac{1}{4}e^2B^2(x^2 + y^2) + p_x^2 + p_y^2 \\ &\qquad + eB(yp_x - xp_y) + eB\hbar\}. \end{aligned} \tag{6}$$

In the last step we have used the fact that any x operator commutes with any y operator, and the commutator values

$$[p_x, x] = [p_y, y] = \frac{\hbar}{\mathrm{i}}. \tag{7}$$

Comparison of (5) and (6), together with the relation $\omega = eB/m$, gives the result

$$bb^\dagger = \frac{H}{\hbar\omega} + \tfrac{1}{2}. \tag{8}$$

The relation

$$b^\dagger b = \frac{H}{\hbar\omega} - \tfrac{1}{2} \tag{9}$$

is obtained in the same way.

(c) Eqs. (8) and (9) show that the operators b and $b^\dagger$, have the same relation to the Hamiltonian as do a and $a^\dagger$, the annihilation and creation

operators for the one-dimensional harmonic oscillator – see (4.18) p. 54.
These relations lead to the result that the energy of the particle is

$$E = (n + \tfrac{1}{2})\hbar\omega, \tag{10}$$

where n is zero or a positive integer.

Comment

The results of this problem apply to a gas of free electrons in a magnetic field. The important feature is that energy is quantised, the energy levels being known as *Landau levels*. This quantisation forms the basis of the *de Haas–van Alphen effect*, which is the oscillation of the magnetic moment of a metal as a function of an applied static magnetic field **B**. The total energy U of the system of electrons depends on the filling of the Landau levels, and the magnetic moment $\mu = -\partial U/\partial B$ oscillates with B as successive orbits are filled.

The electrons in a metal are not free – their wave functions are modified by the periodic potential of the crystal lattice. So the simple theory of the problem has to be modified. But the basic result of quantised energy levels leading to oscillatory behaviour of the magnetisation still holds. Measurement of the variation of μ with B, has provided a great deal of information on the geometry of the Fermi surface. The de Haas–van Alphen effect is discussed in most textbooks on solid state physics – see for example Kittel (1986).

10.4 (a) If $\psi(x,0) = \psi$ is an eigenfunction of the operator a, with eigenvalue μ, then

$$a\psi = \mu\psi. \tag{1}$$

Put

$$\psi = \sum_{n=0}^{\infty} c_n u_n. \tag{2}$$

Since a is the annihilation operator, $au_n = \sqrt{n}\,u_{n-1}$. Therefore, from (1) and (2),

$$c_1 = \mu c_0, \quad c_2 = \frac{\mu}{\sqrt{2}}c_1 = \frac{\mu^2}{\sqrt{(2!)}}c_0, \quad c_3 = \frac{\mu}{\sqrt{3}}c_2 = \frac{\mu^3}{\sqrt{(3!)}}c_0, \tag{3}$$

and so on. Thus

$$c_n = \frac{\mu^n}{\sqrt{(n!)}}\,c_0, \tag{4}$$

giving

$$\psi(x,0) = c_0 \sum_{n=0}^{\infty} \frac{\mu^n}{\sqrt{(n!)}} u_n. \tag{5}$$

(b) The function $\psi(x,t)$ is obtained from $\psi(x,0)$ by multiplying each term in (5) by

$$\exp(-\mathrm{i}E_n t/\hbar) = \exp\{-\mathrm{i}(n+\tfrac{1}{2})\omega t\}. \tag{6}$$

Thus

$$\psi(x,t) = c_0 \sum_{n=0}^{\infty} \frac{\mu^n}{\sqrt{(n!)}} u_n \exp\{-\mathrm{i}(n+\tfrac{1}{2})\omega t\} \tag{7}$$

$$= c_0 \exp(-\tfrac{1}{2}\mathrm{i}\omega t) \sum_{n=0}^{\infty} \frac{\{\mu \exp(-\mathrm{i}\omega t)\}^n}{\sqrt{(n!)}} u_n. \tag{8}$$

Comparison of (5) and (8) shows that, apart from an irrelevant phase factor, they have the same form with μ replaced by $\mu \exp(-\mathrm{i}\omega t)$. So $\psi(x,t)$ is an eigenfunction of the operator a, with eigenvalue $\mu \exp(-\mathrm{i}\omega t)$.

(c) The operator a is defined in terms of position x, and momentum $p_x = (\hbar/\mathrm{i})\,\mathrm{d}/\mathrm{d}x$, by

$$a = \frac{1}{\sqrt{(2m\hbar\omega)}}(m\omega x + \mathrm{i}p_x) = \frac{1}{2\sigma}x + \sigma\frac{\mathrm{d}}{\mathrm{d}x}, \tag{9}$$

where

$$\sigma^2 = \frac{\hbar}{2m\omega}. \tag{10}$$

Putting $\psi = \psi(x,t)$, we have

$$a\psi = \gamma\psi, \tag{11}$$

where

$$\gamma = \mu \exp(-\mathrm{i}\omega t) = \lambda \exp(\mathrm{i}\theta), \quad \theta = \rho - \omega t. \tag{12}$$

From (9) and (11)

$$\frac{\mathrm{d}\psi}{\mathrm{d}x} = \frac{\gamma}{\sigma}\psi - \frac{x}{2\sigma^2}\psi, \tag{13}$$

the solution of which is

$$\psi = c \exp\left(-\frac{x^2}{4\sigma^2} + \frac{\gamma}{\sigma}x\right), \tag{14}$$

where c is a quantity that does not depend on x. Put

$$|\psi|^2 = |c|^2 \exp(-G), \tag{15}$$

where

$$G = \frac{x^2}{2\sigma^2} - \frac{x}{\sigma}(\gamma + \gamma^*) = \frac{x^2}{2\sigma^2} - \frac{2\lambda x}{\sigma}\cos\theta$$

$$= \frac{1}{2\sigma^2}(x - 2\sigma\lambda\cos\theta)^2 - 2\lambda^2\cos^2\theta. \tag{16}$$

Thus

$$|\psi|^2 = |c|^2 \exp(2\lambda^2\cos^2\theta)\exp\{-(x - x_0)^2/2\sigma^2\}, \tag{17}$$

where

$$x_0 = 2\sigma\lambda\cos\theta = 2\sigma\lambda\cos(\rho - \omega t). \tag{18}$$

We finally use the normalisation condition

$$\int_{-\infty}^{\infty} |\psi|^2 \, \mathrm{d}x = 1, \tag{19}$$

together with the result

$$\int_{-\infty}^{\infty} \exp\{-(x - x_0)^2/2\sigma^2\}\, \mathrm{d}x = \sqrt{(2\pi)}\sigma, \tag{20}$$

to obtain

$$|\psi(x, t)|^2 = \frac{1}{\sqrt{(2\pi)}\sigma}\exp\{-(x - x_0)^2/2\sigma^2\}. \tag{21}$$

Note that the factor $\exp(2\lambda^2\cos^2\theta)$ disappears in the normalisation process, i.e.

$$|c|^2 \exp(2\lambda^2\cos^2\theta) = \frac{1}{\sqrt{(2\pi)}\sigma}. \tag{22}$$

(d) The results in (18) and (21) show that $|\psi(x, t)|^2$ is a Gaussian function which undergoes simple harmonic motion with unchanged shape. The standard deviation or width of the function is σ, while the amplitude of the motion is $2\sigma\lambda$, where λ is the modulus of the eigenvalue. The motion is sketched in Fig. 10.2.

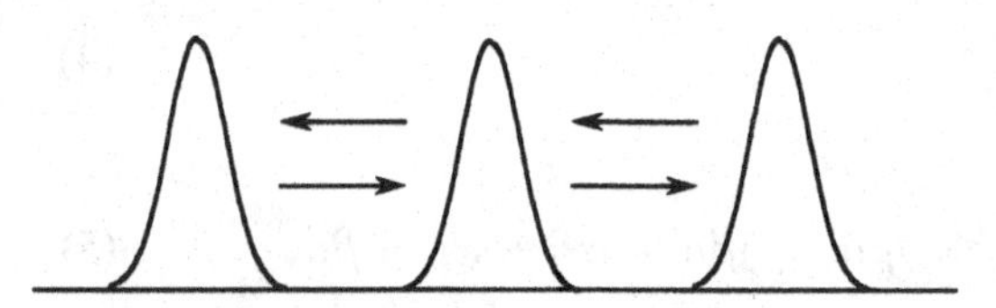

Fig. 10.2. *Motion of coherent state. The Gaussian wave packet undergoes harmonic motion with unchanged shape.*

Comments

(1) The eigenfunction $\psi(x, t)$ of the lowering operator a is known as a *coherent state*. The operator a does not correspond to a physical observable and is not Hermitian. Consequently its eigenvalues are in general complex. A further consequence is that the eigenfunctions corresponding to different eigenvalues μ and μ' are not orthogonal. As can be seen from the problem *any* complex number is a possible eigenvalue.

(2) Coherent states have important applications in quantum electronics and quantum optics. The states u_n, the eigenfunctions of the Hamiltonian for the harmonic oscillator, are useful for characterising states in electrodynamical systems involving a small, fixed number, of photons, but in the light beam of a laser, say, we are dealing with states in which the number of photons is large and intrinsically uncertain. In such cases, the coherent states $\psi(x, t)$ provide a good description of the role played by the photons in the wavelike properties of the radiation field of the laser. For further discussion of coherent states and their physical interpretation, see Schubert and Wilhelmi (1986) and Glauber (1963).

10.5 (a) The state χ_j may be written as the product function

$$\chi_j = \alpha\alpha\alpha \dots \alpha\beta\alpha \dots \alpha\alpha \tag{1}$$

$$i \rightarrow 1\,2 \qquad\qquad j \qquad\qquad \mathrm{N}$$

Writing the scalar product in $\boldsymbol{\sigma}_i \cdot \boldsymbol{\sigma}_{i+1}$ in terms of the x, y, z components we have

$$\boldsymbol{\sigma}_i \cdot \boldsymbol{\sigma}_{i+1} = \sigma_{ix}\sigma_{i+1,x} + \sigma_{iy}\sigma_{i+1,y} + \sigma_{iz}\sigma_{i+1,z}. \tag{2}$$

$\boldsymbol{\sigma}_i \cdot \boldsymbol{\sigma}_{i+1}$ operates only on the states in (1) corresponding to the i and $i + 1$ ions. It treats the other states in the product as constants.

If j does not equal i or $i + 1$, we have, using the results in (5.3.6),

$$(\sigma_{ix}\sigma_{i+1,x} + \sigma_{iy}\sigma_{i+1,y} + \sigma_{iz}\sigma_{i+1,z})\alpha\alpha = \beta\beta - \beta\beta + \alpha\alpha = \alpha\alpha, \tag{3}$$

where each $\alpha\alpha$ or $\beta\beta$ refers to the i and $i + 1$ ions. Thus

$$\boldsymbol{\sigma}_i \cdot \boldsymbol{\sigma}_{i+1}\chi_j = \chi_j. \tag{4}$$

If $j = i$, we have

$$(\sigma_{ix}\sigma_{i+1,x} + \sigma_{iy}\sigma_{i+1,y} + \sigma_{iz}\sigma_{i+1,z})\beta\alpha = \alpha\beta + \alpha\beta - \beta\alpha. \tag{5}$$

Thus

$$\boldsymbol{\sigma}_i \cdot \boldsymbol{\sigma}_{i+1}\chi_i = 2\chi_{i+1} - \chi_i. \tag{6}$$

If $j = i + 1$, we have

$$(\sigma_{ix}\sigma_{i+1,x} + \sigma_{iy}\sigma_{i+1,y} + \sigma_{iz}\sigma_{i+1,z})\alpha\beta = \beta\alpha + \beta\alpha - \alpha\beta. \tag{7}$$

Thus

$$\boldsymbol{\sigma}_i \cdot \boldsymbol{\sigma}_{i+1}\chi_{i+1} = 2\chi_i - \chi_{i+1}. \tag{8}$$

Note that a subscript $N + 1$ is to be interpreted as 1.

(b) Consider $(\sum_{i=1}^{N}\boldsymbol{\sigma}_i \cdot \boldsymbol{\sigma}_{i+1})\chi_n$. There are $N - 2$ terms in the sum where neither i nor $i + 1$ is equal to n. Each of these terms is equal to χ_n. The other two terms are given by (6) and (8). Thus

$$\left(\sum_{i=1}^{N}\boldsymbol{\sigma}_i \cdot \boldsymbol{\sigma}_{i+1}\right)\chi_n = N\chi_n + 2(\chi_{n-1} + \chi_{n+1} - 2\chi_n), \tag{9}$$

whence

$$\begin{aligned} H\chi &= -\tfrac{1}{2}J\left(\sum_{i=1}^{N}\boldsymbol{\sigma}_i \cdot \boldsymbol{\sigma}_{i+1}\right)\sum_{n=1}^{N} c_n\chi_n \\ &= -\tfrac{1}{2}JN\sum_{n=1}^{N} c_n\chi_n + J\sum_{n=1}^{N} c_n(2\chi_n - \chi_{n-1} - \chi_{n+1}) \\ &= E\sum_{n=1}^{N} c_n\chi_n. \end{aligned} \tag{10}$$

Equate terms in χ_n, and put $E_0 = -\tfrac{1}{2}JN$. This gives

$$(E - E_0)c_n = J(2c_n - c_{n+1} - c_{n-1}). \tag{11}$$

(c) If

$$c_n = \frac{1}{\sqrt{N}}\exp(iqna), \tag{12}$$

then

$$c_{n+1} = \exp(iqa)c_n, \quad \text{and} \quad c_{n-1} = \exp(-iqa)c_n. \tag{13}$$

Inserting these relations in (11) we obtain

$$E - E_0 = 2J(1 - \cos qa). \tag{14}$$

Comments

(1) It is readily shown that E_0 is the energy of the ground state χ_0, in which all the ions are in the state α. From (4)

$$\boldsymbol{\sigma}_i \cdot \boldsymbol{\sigma}_{i+1}\chi_0 = \chi_0. \tag{15}$$

The relation holds for all values of i. Therefore

$$-\tfrac{1}{2}J\left(\sum_{i=1}^{N}\boldsymbol{\sigma}_i\cdot\boldsymbol{\sigma}_{i+1}\right)\chi_0 = -\tfrac{1}{2}JN\chi_0 = E_0\chi_0. \tag{16}$$

(2) The magnetic excitation represented by the state $\chi = \Sigma c_n\chi_n$, where the c_n are given by (12), is known as a *spin wave*. Putting $E - E_0 = \hbar\omega$, we have

$$\hbar\omega = 2J(1 - \cos qa). \tag{17}$$

The relation between ω and the wavenumber q is known as the *dispersion relation* of the spin wave. It is plotted in Fig. 10.3. For $qa \ll 1$, $\hbar\omega \rightarrow Jq^2a^2$. The quadratic dependence of ω on q, at low q, is a general result for a spin wave. It is to be contrasted with the linear dependence for phonons.

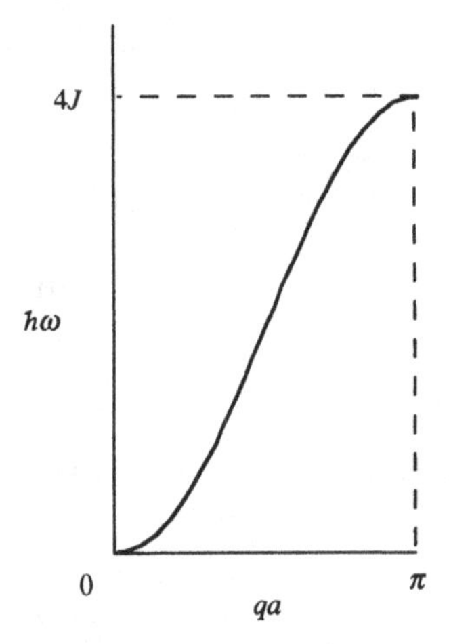

Fig. 10.3. *Spin-wave dispersion relation.*

The energy of the ground state is of the order of $-JN$, whereas the energy of the spin-wave excitation above the ground state is of the order of J. So for large N, which is normally the case in a physical system, a spin wave has relatively low energy.

(3) If there were no interactions between the ions, a state such as χ_j, in which the spin of the jth ion is reversed, would be a stationary state. If there is an interaction, this is not so. When a single spin is reversed, the interaction tends to reverse the spin of a neighbour, with the first ion going back to the unreversed state. So the spin wave can be imagined as a wave representing the probability of a reversed spin, moving through the

lattice of ions with phase velocity ω/q, where q is the wave-number of the wave, and ω is given by (17).

(4) The values of q do not extend indefinitely, because two values q and q', such that $q'a = qa + 2\pi n'$, where n' is an integer, give the same value of c_n at the site of the nth ion. This follows from the result

$$\exp(\mathrm{i}q'na) = \exp(\mathrm{i}qna)\exp(2\pi\mathrm{i}nn') = \exp(\mathrm{i}qna) \tag{18}$$

for all values of n. We normally select q to have the smallest range of values, i.e. to lie in the range $-\pi/a < q < \pi/a$, which is the one-dimensional counterpart of the 1st Brillouin zone.

The fact that we have a ring of ions means that the values of q are discrete. The spin wave must have an integral number n'' of wavelengths round the ring. So

$$q = \frac{2\pi}{\lambda} = \frac{2\pi n''}{Na}. \tag{19}$$

Since the range of q is $2\pi/a$, there are N values for n'', i.e. there are N independent spin waves. This is to be expected as there are N ions in the ring with N independent spin states.

We have chosen a ring of ions for the problem, rather than a linear chain, in order to avoid complications with boundary conditions. In practice, the effects at the ends of the chain are usually negligible. Moreover, in the latter case it is conventional (if not very physical) to impose the boundary condition that the spin waves are periodic in the length of the chain, which is mathematically equivalent to the condition for the ring.

10.6 (a) The matrix elements are given by

$$\langle 1|V|1\rangle = \int_{\substack{\text{all}\\ \text{space}}} (R(r)Y_{11})^* V(x,y,z)R(r)Y_{11}\,\mathrm{d}\tau, \tag{1}$$

and so on. (The two 1s in $\langle 1|V|1\rangle$ refer to the m values of the spherical harmonics.) Due to the form of the function V, it is convenient to express the spherical harmonics Y_{11}, Y_{10}, Y_{1-1} in terms of Cartesian coordinates

$$x = r\sin\theta\cos\phi, \quad y = r\sin\theta\sin\phi, \quad z = r\cos\theta. \tag{2}$$

From Table 4.1, the spherical harmonics for $l = 1$ are

$$Y_{11} = -(3/8\pi)^{1/2}(x + \mathrm{i}y)/r, \quad Y_{10} = (3/4\pi)^{1/2}z/r,$$
$$Y_{1-1} = (3/8\pi)^{1/2}(x - \mathrm{i}y)/r. \tag{3}$$

So

$$\langle 1|V|1\rangle = \gamma_1 \int_{\substack{\text{all}\\ \text{space}}} \left|\frac{R(r)}{r}\right|^2 G_{11}\{Ax^2 + By^2 - (A+B)z^2\}\,\mathrm{d}\tau, \qquad (4)$$

where γ_1 is a constant, and

$$G_{11} = (x + \mathrm{i}y)^*(x + \mathrm{i}y) = x^2 + y^2. \qquad (5)$$

Now

$$\begin{aligned}(x^2 + y^2)\{Ax^2 + By^2 - (A+B)z^2\} \\ = A(x^4 + x^2y^2 - x^2z^2 - y^2z^2) \\ + B(y^4 + x^2y^2 - x^2z^2 - y^2z^2).\end{aligned} \qquad (6)$$

If $f(r)$ is a function that depends only on the magnitude of r, then

$$\int f(r)x^4\,\mathrm{d}\tau = \int f(r)y^4\,\mathrm{d}\tau, \qquad (7)$$

because the integration is taken over all space, and the integrands differ only in their x, y labelling. Similarly

$$\int f(r)x^2y^2\,\mathrm{d}\tau = \int f(r)y^2z^2\,\mathrm{d}\tau = \int f(r)z^2x^2\,\mathrm{d}\tau. \qquad (8)$$

The results in (4) to (8) show that the matrix element has the form

$$\langle 1|V|1\rangle = \gamma(A + B), \qquad (9)$$

where γ is a constant.

Now consider $\langle 1|V|-1\rangle$. It is the same as the expression for $\langle 1|V|1\rangle$ in (4), but with G_{11} replaced by

$$G_{1-1} = -(x + \mathrm{i}y)^*(x - \mathrm{i}y) = -x^2 + y^2 + 2\mathrm{i}xy \qquad (10)$$

– see (3). The term in xy gives zero contribution to the integral, because the overall integrand for the term is an odd function of x and of y, which gives zero when integrated over all space. From the previous reasoning

$$\langle 1|V|-1\rangle = -\gamma(A - B). \qquad (11)$$

Similarly

$$\langle -1|V|1\rangle = -\gamma(A - B), \qquad (12)$$

$$\langle -1|V|-1\rangle = \gamma(A + B). \qquad (13)$$

The expression for $\langle 0|V|0\rangle$ is the same as (4) with G_{11} replaced by $G_{00} = 2z^2$. Thus

$$\langle 0|V|0\rangle = -2\gamma(A + B). \qquad (14)$$

The matrix elements between the $m = 0$ and the $m = \pm 1$ states are zero, because the integrands consist of terms which are odd functions of x, y, z. The matrix is symmetric, and is thus given by

$$\gamma \begin{bmatrix} A + B & 0 & -A + B \\ 0 & -2(A + B) & 0 \\ -A + B & 0 & A + B \end{bmatrix}. \tag{15}$$

(b) To calculate the matrix of V on the basis of the states $M_J = 3/2, 1/2, -1/2, -3/2$ we need the results in the second part of Table 5.1, p. 96. In calculating the matrix elements of V we use the fact that V is a function of space coordinates, and not of spin. Since

$$\langle \alpha | \beta \rangle = \langle \beta | \alpha \rangle = 0, \tag{16}$$

we need consider only terms in which the two spin functions are the same. For such terms

$$\langle \alpha | \alpha \rangle = \langle \beta | \beta \rangle = 1. \tag{17}$$

Thus

$$\langle 3/2 | V | 3/2 \rangle = \langle 1\alpha | V | 1\alpha \rangle = \langle 1 | V | 1 \rangle = \gamma(A + B), \tag{18}$$

$$\langle 3/2 | V | 1/2 \rangle = \sqrt{\tfrac{2}{3}} \langle 1\alpha | V | 0\alpha \rangle + \sqrt{\tfrac{1}{3}} \langle 1\alpha | V | 1\beta \rangle = 0. \tag{19}$$

The first term on the right-hand side of (19) is zero because $\langle 1 | V | 0 \rangle = 0$, and the second term is zero because the two spin functions are different. Continuing in the same way, with the results in (15) and Table 5.1, we obtain the required matrix

$$\begin{bmatrix} a & 0 & b & 0 \\ 0 & -a & 0 & b \\ b & 0 & -a & 0 \\ 0 & b & 0 & a \end{bmatrix}, \tag{20}$$

where $a = \gamma(A + B)$, and $b = \gamma(B - A)/\sqrt{3}$.

(c) The four M_J states are degenerate for the free ion. Hence the first-order change in the energy of the ground state is given by the eigenfunctions of the matrix in (20), and the eigenstates are the corresponding eigenvectors. The calculation is most easily done if the order of the basis states is changed from $M_J = 3/2, 1/2, -1/2, -3/2$ to $M_J = 3/2, -1/2, -3/2, 1/2$. The matrix then becomes

$$\begin{bmatrix} a & b & 0 & 0 \\ b & -a & 0 & 0 \\ 0 & 0 & a & b \\ 0 & 0 & b & -a \end{bmatrix}, \tag{21}$$

which separates into two 2×2 matrices. The eigenvalues λ, and the values of p and q for the eigenvectors $p|3/2\rangle + q|-1/2\rangle$, of the first matrix, are given by the equations

$$\begin{bmatrix} a & b \\ b & -a \end{bmatrix}\begin{bmatrix} p \\ q \end{bmatrix} = \lambda\begin{bmatrix} p \\ q \end{bmatrix}, \tag{22}$$

with solutions

$$\lambda = \pm\surd(a^2 + b^2), \quad \frac{p}{q} = \frac{\lambda + a}{b}. \tag{23}$$

If we define θ by $\tan\theta = b/a$, we have, for the eigenvalue $\lambda = +\surd(a^2 + b^2)$,

$$\frac{p}{q} = \frac{\surd(a^2 + b^2) + a}{b} = \frac{1 + \cos\theta}{\sin\theta} = \frac{\cos(\theta/2)}{\sin(\theta/2)}. \tag{24}$$

The normalisation condition, $p^2 + q^2 = 1$, then gives

$$p = \cos(\theta/2), \quad q = \sin(\theta/2). \tag{25}$$

The eigenvector for the eigenvalue $\lambda = -\surd(a^2 + b^2)$ is most readily obtained from the orthogonality condition. Thus

$$p = \sin(\theta/2), \quad q = -\cos(\theta/2). \tag{26}$$

The other 2×2 matrix is identical, so it has the same pair of eigenvalues, and the same pair of values for p and q, which give the eigenvectors in terms of the states $|-3/2\rangle$ and $|1/2\rangle$. The splitting caused by the crystal field is shown in Fig. 10.4.

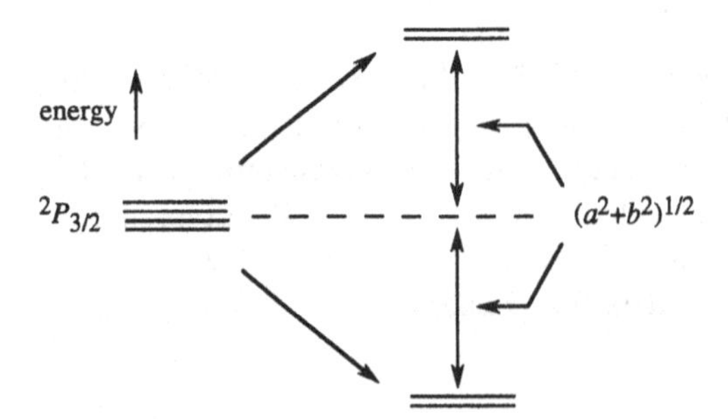

Fig. 10.4. *Splitting of a $^2P_{3/2}$ level due to an orthorhombic crystal field. The four-fold degenerate $^2P_{3/2}$ level is split into two doubly-degenerate levels.*

Comments

(1) This problem illustrates how the electrostatic field of a crystal lattice splits a degenerate energy level of a free ion. The effect is known as *crystal–field splitting*. The electrostatic potential has the symmetry of the lattice. We have shown that the field for an orthorhombic lattice splits the four-fold degenerate energy of a free $^2P_{3/2}$ ion into two doubly-degenerate energy levels.

For a tetragonal lattice, $A = B$ in the expression for $V(\mathbf{r})$. Thus $b = 0$ in (20). The unperturbed energy is again split into two doubly degenerate levels, with energies $\lambda = \pm a$. Since $b = 0$, $\theta = 0$. So the eigenfunctions are $|3/2\rangle$ and $|-3/2\rangle$ for $\lambda = a$, and $|1/2\rangle$ and $|-1/2\rangle$ for $\lambda = -a$.

(2) At all points in space, other than the origin, the potential $V(\mathbf{r})$ must satisfy Laplace's equation

$$\nabla^2 V = \frac{\partial^2 V}{\partial^2 x} + \frac{\partial^2 V}{\partial^2 y} + \frac{\partial^2 V}{\partial^2 z} = 0. \tag{27}$$

A potential with orthorhombic symmetry has the form

$$V(\mathbf{r}) = Ax^2 + By^2 + Cz^2, \tag{28}$$

where A, B, C are constants. To satisfy (27) we must have

$$A + B + C = 0, \tag{29}$$

which accounts for the form of $V(\mathbf{r})$ in the problem.

For a lattice with cubic symmetry, we have

$$A = B = C, \tag{30}$$

which means, in view of (29), that $A = B = C = 0$. Therefore, in this case, a cubic lattice would produce no splitting of the ground-state energy of the free ion. The theory shows that, for an ion in a P state ($L = 1$), even though the expression for V may contain terms in higher powers of x, y, z, it remains true that there is no splitting. However, for higher values of L, terms in x^4, etc. in $V(\mathbf{r})$ may give rise to splitting by the field of a cubic lattice.

(3) There is a theorem due to Kramers which states that, for a system with an odd number of electrons, then, whatever the symmetry of the crystal field, at least a two-fold degeneracy will remain. In the present problem the half-integral value of J shows that the number of electrons is odd. So the result that the levels are doubly degenerate is in accordance with Kramers' theorem.

10.7 (a) The operator corresponding to the z component of the magnetic dipole moment of the proton is $\gamma_p(\hbar/2)\sigma_z$. Therefore, the term in the Hamiltonian due to the magnetic field **B** along z is

$$H^{(0)} = -\gamma_p\frac{\hbar}{2}B\sigma_z = -\frac{\hbar}{2}\omega_L\sigma_z. \tag{1}$$

The term in the Hamiltonian due to the rotating field is

$$\begin{aligned} H^{(1)} &= -\gamma_p\frac{\hbar}{2}B_r(\sigma_x\cos\omega t - \sigma_y\sin\omega t) \\ &= -\frac{\hbar}{2}\omega_r(\sigma_x\cos\omega t - \sigma_y\sin\omega t). \end{aligned} \tag{2}$$

(b) The eigenfunctions of $H^{(0)}$ are α and β, with energies

$$E_\alpha = -\frac{\hbar}{2}\omega_L, \quad E_\beta = \frac{\hbar}{2}\omega_L. \tag{3}$$

Express the state function as

$$\psi = c_\alpha\exp(\mathrm{i}\omega_L t/2)\alpha + c_\beta\exp(-\mathrm{i}\omega_L t/2)\beta, \tag{4}$$

where c_α and c_β are time-varying coefficients. The time variation of ψ is given by

$$\frac{\partial\psi}{\partial t} = \frac{1}{\mathrm{i}\hbar}H\psi, \tag{5}$$

where

$$H = H^{(0)} + H^{(1)} = -\frac{\hbar}{2}\{\omega_L\sigma_z + \omega_r(\cos\omega t\,\sigma_x - \sin\omega t\,\sigma_y)\}. \tag{6}$$

The relations

$$\begin{aligned} &\sigma_x\alpha = \beta, \quad \sigma_y\alpha = \mathrm{i}\beta, \quad \sigma_z\alpha = \alpha, \\ &\sigma_x\beta = \alpha, \quad \sigma_y\beta = -\mathrm{i}\alpha, \quad \sigma_z\beta = -\beta, \end{aligned} \tag{7}$$

give

$$\begin{aligned} H\alpha &= -\frac{\hbar}{2}\{\omega_L\alpha + \omega_r\exp(-\mathrm{i}\omega t)\beta\}, \\ H\beta &= -\frac{\hbar}{2}\{-\omega_L\beta + \omega_r\exp(\mathrm{i}\omega t)\alpha\}. \end{aligned} \tag{8}$$

Substitute (4) in (5), using (8), and equate the coefficients of α and β. This gives

$$\dot{c}_\alpha = \tfrac{1}{2}\mathrm{i}\omega_r\exp(\mathrm{i}qt)c_\beta, \tag{9}$$

$$\dot{c}_\beta = \tfrac{1}{2}\mathrm{i}\omega_\mathrm{r}\exp(-\mathrm{i}qt)c_\alpha, \tag{10}$$

where

$$q = \omega - \omega_\mathrm{L}. \tag{11}$$

(c) To obtain a differential equation for c_β, we differentiate (10) with respect to time, and eliminate c_α and $\dot{c}_\alpha$ in the resulting expression by means of (9) and (10). This gives

$$\ddot{c}_\beta + \mathrm{i}q\dot{c}_\beta + \tfrac{1}{4}\omega_\mathrm{r}^2 c_\beta = 0, \tag{12}$$

with solution

$$c_\beta = \exp[\mathrm{i}\{-q \pm \sqrt{(q^2 + \omega_\mathrm{r}^2)}\}t/2]. \tag{13}$$

The general solution of (12) is

$$c_\beta = \exp(-\mathrm{i}qt/2)[A\exp\{\mathrm{i}\sqrt{(q^2 + \omega_\mathrm{r}^2)}t/2\} + B\exp\{-\mathrm{i}\sqrt{(q^2 + \omega_\mathrm{r}^2)}t/2\}], \tag{14}$$

where A and B are constants fixed by the initial conditions. These are that at $t = 0$

$$c_\alpha = 1, \quad c_\beta = 0, \quad \dot{c}_\beta = \tfrac{1}{2}\mathrm{i}\omega_\mathrm{r}. \tag{15}$$

The last relation comes from (10) at $t = 0$. These relations give

$$A = -B = \frac{\omega_\mathrm{r}}{2\sqrt{(q^2 + \omega_\mathrm{r}^2)}}, \tag{16}$$

whence

$$c_\beta = \frac{\mathrm{i}\omega_\mathrm{r}}{\sqrt{(q^2 + \omega_\mathrm{r}^2)}}\exp(-\mathrm{i}qt/2)\sin\{\sqrt{(q^2 + \omega_\mathrm{r}^2)}t/2\}, \tag{17}$$

i.e.

$$|c_\beta|^2 = \frac{\omega_\mathrm{r}^2}{q^2 + \omega_\mathrm{r}^2}\sin^2\{\sqrt{(q^2 + \omega_\mathrm{r}^2)}t/2\}. \tag{18}$$

(d) Eq. (10) is exact. It gives the value of $\dot{c}_\beta$ at time t in terms of the value of c_α at time t. In first-order perturbation theory, c_α is put equal to 1, its value at $t = 0$. The equation becomes

$$\dot{c}_\beta = \tfrac{1}{2}\mathrm{i}\omega_\mathrm{r}\exp(-\mathrm{i}qt), \tag{19}$$

with solution

$$c_\beta = -\frac{\omega_\mathrm{r}}{2q}\{\exp(-\mathrm{i}qt) - 1\}, \tag{20}$$

whence

$$|c_\beta|^2 = \frac{\omega_r^2}{q^2}\sin^2\frac{qt}{2}. \tag{21}$$

Comparing this with the exact expression in (18), we see it is a good approximation, provided $q \gg \omega_r$. This is to be expected, because, if $q \gg \omega_r$, then $c_\beta \ll 1$, which means (since $c_\alpha^2 + c_\beta^2 = 1$) that c_α is only slightly smaller than 1. Therefore, replacing it by 1 in (10) is a good approximation.

Comments

(1) The phenomenon of spin reversal brought about by an oscillating magnetic field is known as *magnetic resonance*. For a fixed value of ω, $|c_\beta|^2$ oscillates with time, between zero and

$$M = \frac{\omega_r^2}{(q^2 + \omega_r^2)} = \frac{\omega_r^2}{\{(\omega - \omega_L)^2 + \omega_r^2\}}. \tag{22}$$

M is plotted as a function of ω in Fig. 10.5. It has a maximum when

$$\omega = \omega_L, \tag{23}$$

and drops to half its maximum value when

$$\omega - \omega_L = \pm\omega_r, \tag{24}$$

i.e. the half-width of the curve at half-height is ω_r. From (3) the resonance condition in (23) is equivalent to

$$\hbar\omega = E_\beta - E_\alpha. \tag{25}$$

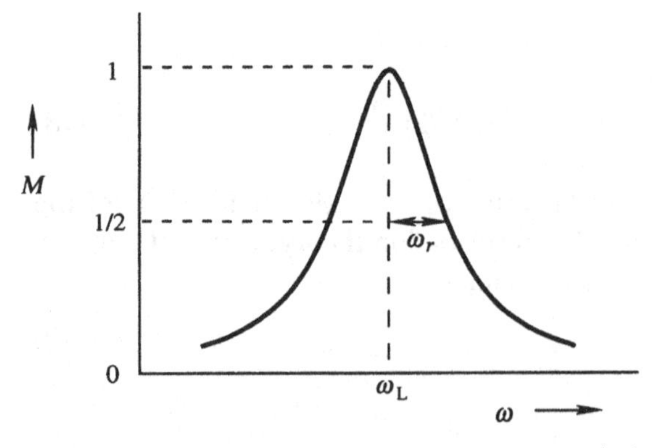

Fig. 10.5. *Magnetic resonance – the probability $|c_\beta|^2$ of a spin transition due to a rotating magnetic field B_r varies sinusoidally with time. The diagram shows the maximum value M of $|c_\beta|^2$ as a function of the angular frequency of B_r.*

The relative width of the resonance function in Fig. 10.5 is

$$\frac{\omega_r}{\omega_L} = \frac{B_r}{B}. \tag{26}$$

So the smaller the amplitude of the rotating field B_r compared to the value of the static magnetic field along z, the sharper the resonance. The time variation of $|c_\beta|^2$ is very rapid and is usually not observed. So the $\sin^2$ term in (18) may be replaced by its time-averaged value of $\frac{1}{2}$.

(2) If the initial state is β, i.e. the condition at $t = 0$ is $c_\alpha = 0$, $c_\beta = 1$, the calculation is the same, and the expression for $|c_\alpha|^2$ is identical to (18). This is an example of the principle of detailed balancing – p. 173.

(3) The principle of magnetic resonance has many applications, for example, it provides the most accurate method of measuring a static magnetic field. A system of protons, such as a small sample of water, is placed in the region where the magnetic field B is to be determined. A coil, through which a radio-frequency current may be passed, is placed round the sample, and the frequency ω of the current through the coil is varied until the resonance condition is detected. If there were exactly equal numbers of protons in the states α and β, there would be no effect, even at resonance. This is because, by the principle of detailed balancing, there would be equal numbers of transitions $\alpha \to \beta$, and $\beta \to \alpha$. However, in the presence of $\mathbf{B}$, the protons in the state α have lower energy than those in state β. Therefore the Boltzmann factor results in the number of protons in state α being slightly larger than those in state β. The number of transitions per unit time is equal to the product of the number of particles in the initial state and the probability of a transition. So the number of transitions $\alpha \to \beta$ exceeds that for the reverse direction, and energy flows from the radio-frequency generator into the proton system, a condition which may be detected.

The value of ω at resonance gives the value of B from the relation

$$\omega = \omega_L = \gamma_p B. \tag{27}$$

To convert the frequency into a value of B, we need only the value of γ_p, which is known with a fractional error of about 3×10^{-7}. The method provides a direct and accurate method of determining B.

(4) Another use of magnetic resonance is to change the spin of a beam of polarised neutrons. The neutrons pass through a device with a static magnetic field either parallel or antiparallel to the direction of the spin, and a rotating field in a perpendicular direction. In this case it is the time-varying form of $|c_\beta|^2$ in (18) that is relevant. The time that the neutrons spend in the magnetic field region is known from their velocity and the geometry of the device. The value of the rotating field B_r is

adjusted so that the sine term in (18) is unity. Thus, at resonance, $|c_\beta|^2 = 1$ (or would be if all the experimental conditions were perfect). The device is known as a *radio-frequency spin flipper*.

(5) In practice, the oscillatory magnetic field in magnetic resonance does not rotate, but remains in a fixed direction in the xy plane. However, this is equivalent to two fields rotating in opposite directions in the plane. The one rotating from y to x (I in Fig. 10.6) is the one we have defined in the problem. When $\omega = \omega_L$, it is this one that brings about transitions for protons. The other field has negative ω and brings about negligible transitions.

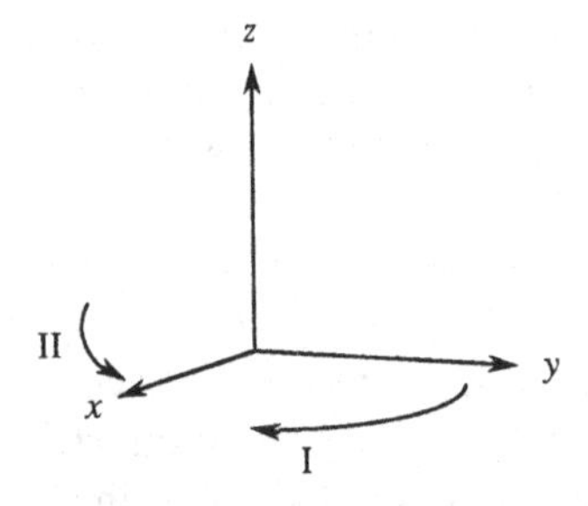

Fig. 10.6. *An oscillating magnetic field along x is equivalent to the rotating magnetic fields* I *plus* II.

We can obtain a physical picture of the quantum mechanical results. In the absence of the oscillatory field in the xy plane, the proton remains in the state α or β. The picture for the state α is shown in Fig. 10.7(a). The spin angular momentum vector is at an angle to the z axis, reflecting the fact that the z-component is $\hbar/2$, which is less than $\sqrt{(3/4)}\hbar$, the magnitude of the vector. The field $\mathbf{B}$ produces a couple $\mathbf{G} = \boldsymbol{\mu} \times \mathbf{B}$, where $\boldsymbol{\mu}$ is the magnetic dipole moment of the proton. Because the proton has angular momentum in the same direction as $\boldsymbol{\mu}$, the spin vector precesses in the direction shown in the figure. The angular frequency of the rotation, given by equating the couple to the rate of change of angular momentum, is ω_L. The effect of a rotating magnetic field in the xy plane is to give the proton spin an additional motion, namely, precession about the instantaneous direction of the field. We may interpret this as a probability of bringing about a transition to the state β. This probability will remain small unless the oscillatory field has the same frequency as the original precessional motion, and rotates in the same sense. This explains why it is field I, and not II, in Fig. 10.6 that brings about transitions. You can verify that a proton in the state β precesses about $\mathbf{B}$ in the same sense as α, so the field I also brings about transitions from β to α.

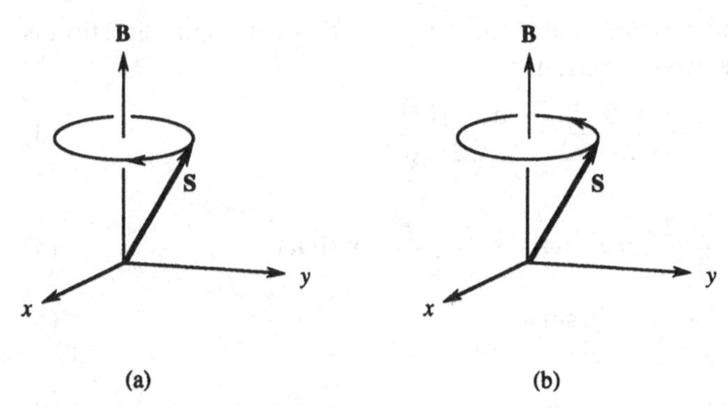

Fig. 10.7. *Precession of spin angular momentum vector* **S** *in magnetic field* **B** *for (a) positive and (b) negative gyromagnetic ratio.*

For the electron and the neutron, the magnetic dipole moment is in the opposite direction to the spin, i.e. the gyromagnetic ratio is negative. For these particles, the sense of the precessional motion is opposite to that of the proton – Fig. 10.7(b) – and it is the rotating field II that brings about transitions.

10.8 (a) The neutron beams and spin directions for zero magnetic field are shown in Fig. 10.8. The calculation of the effect of the magnetic field is the same as that in Problem 8.1 with the x and z axes interchanged. Since the magnetic field is along the x direction, we express the spin wave function in terms of the eigenfunctions of σ_x, which are

$$(\alpha + \beta)/\sqrt{2}, \quad (\alpha - \beta)/\sqrt{2}, \tag{1}$$

with corresponding eigenvalues

$$\mu_n B, \quad -\mu_n B. \tag{2}$$

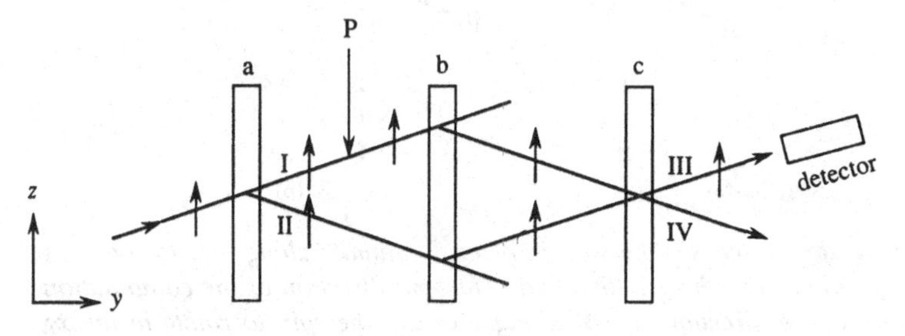

Fig. 10.8. *Beams and spin directions in neutron interferometer for zero magnetic field at* P. *The x axis is perpendicular to the plane of the diagram.*

At $t = 0$, when the neutrons enter the magnetic field, the spin direction is along $+z$, and the wave function is

$$\psi(0) = \alpha = \frac{\alpha + \beta}{\sqrt{2}}\frac{1}{\sqrt{2}} + \frac{\alpha - \beta}{\sqrt{2}}\frac{1}{\sqrt{2}}. \tag{3}$$

Thus

$$\psi(t) = \frac{\alpha + \beta}{2}\exp(-\mathrm{i}\omega t) + \frac{\alpha - \beta}{2}\exp(\mathrm{i}\omega t) \tag{4}$$

$$= \alpha\cos\omega t - \mathrm{i}\beta\sin\omega t, \tag{5}$$

where

$$\hbar\omega = \mu_{\mathrm{n}} B. \tag{6}$$

When $t = \pi\hbar/2\mu_{\mathrm{n}} B$,

$$\omega t = \frac{\pi}{2}, \quad \text{and} \quad \psi(t) = -\mathrm{i}\beta. \tag{7}$$

Since there is no component of α in the state $\psi(t)$, the spin of the beam leaving the magnetic field is along $-z$. See Fig. 10.9(a).

(b) The wave function is expressed in terms of $(\alpha + \mathrm{i}\beta)/\sqrt{2}$ and $(\alpha - \mathrm{i}\beta)/\sqrt{2}$, the eigenfunctions of σ_y. By the same reasoning as above we obtain

$$\psi(t) = \frac{\alpha + \mathrm{i}\beta}{2}\exp(-\mathrm{i}\omega t) + \frac{\alpha - \mathrm{i}\beta}{2}\exp(\mathrm{i}\omega t) \tag{8}$$

$$= \alpha\cos\omega t + \beta\sin\omega t. \tag{9}$$

When $\omega t = \pi/2$, $\psi(t) = \beta$. So the spin is again reversed. See Fig. 10.9(b).

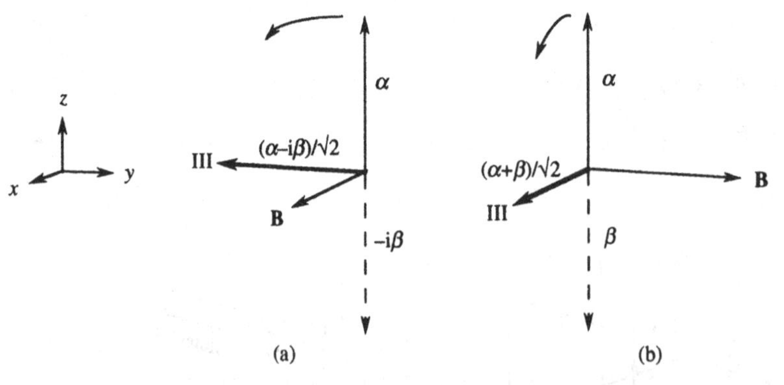

Fig. 10.9. *(a)* **B** *along x causes the spin in beam* I, *initially along* $+z$, *to rotate as shown in the yz plane into the* $-z$ *direction. The spin direction of the combination beam* III *is in the* $-y$ *direction. (b)* **B** *along y causes the spin to rotate in the xz plane into the* $-z$ *direction, but the spin of beam* III *is in the* $+x$ *direction. The spin wave function is given beside the spin direction.*

Although the spin is reversed in each case, the spin wave functions are $-\mathrm{i}\beta$ when the magnetic field is along x, and β when the magnetic field is along y, i.e. there is a phase difference between the two wave functions. The state for the second beam, which does not pass through the magnetic field, is α in both cases. The two beams are finally combined coherently, so the wave function Ψ of the combined beam III – Fig. 10.8 – is, apart from the normalisation factor, simply the sum of the wave functions of the two component beams. Thus

$$\Psi = (\alpha - \mathrm{i}\beta)/\sqrt{2}, \quad \mathbf{B} \text{ along } x, \tag{10}$$

$$\Psi = (\alpha + \beta)/\sqrt{2}, \quad \mathbf{B} \text{ along } y. \tag{11}$$

We see that the first Ψ represents a state with the spin pointing along the $-y$ direction, and the second a state with the spin pointing along $+x$.

These results are illustrated in Fig. 10.9. It may seem strange that although the final direction of the spin is the same whether the magnetic field is along x or along y, the spin state function is different in the two cases. This is a purely quantum mechanical result with no classical counterpart. But we can see from the figure that the results are physically consistent in that, whatever the direction of the field $\mathbf{B}$ in the xy plane, the three directions given by the spin of the combined beam, the magnetic field, and the z axis, form a right-handed set of axes. In other words, the phase factor in the spin wave function of beam I after it has traversed the magnetic field contains the information of the direction of the field which rotated the spin. This information is expressed physically by the spin direction of the coherently combined final beam III. Notice that, if the beams I and II are not combined, no measurement on beam I can reveal the phase factor of the spin function, i.e. tell us the direction of the magnetic field. An experimental demonstration of the coherent combination of two neutron beams has been given by Summhammer *et al*. (1983).

(c) When $t = \pi\hbar/\mu_{\mathrm{n}}B$, then $\omega t = \pi$, and (5) becomes

$$\psi(t) = -\alpha, \tag{12}$$

i.e. the spin has rotated through 2π and points along the initial direction $+z$, but the sign of the state function is reversed. This again is a purely quantum mechanical result, and is generally true for a particle of half-integral spin.

When $t = 2\pi\hbar/\mu_{\mathrm{n}}B$, then $\omega t = 2\pi$, and (5) becomes

$$\psi(t) = \alpha, \tag{13}$$

i.e. it requires a spin rotation of 4π to bring the particle back to its original state function, including the sign. As in the previous section we cannot detect the change in sign following a 2π rotation by any experiment on beam I alone. But the change in sign is revealed by measuring the intensity of the combined beam III. For the 2π rotation, the two beams are out of phase and the intensity of beam III is (ideally) zero, whereas, for the 4π rotation, the two beams are in phase, and beam III has maximum intensity.

Some results by Rauch *et al*. (1975) are shown in Fig. 10.10, where the counting rate in beam III is plotted against the current in the coil producing the magnetic field **B**. The distance between the maxima agrees with the calculated change in the current for a 4π rotation of the spin.

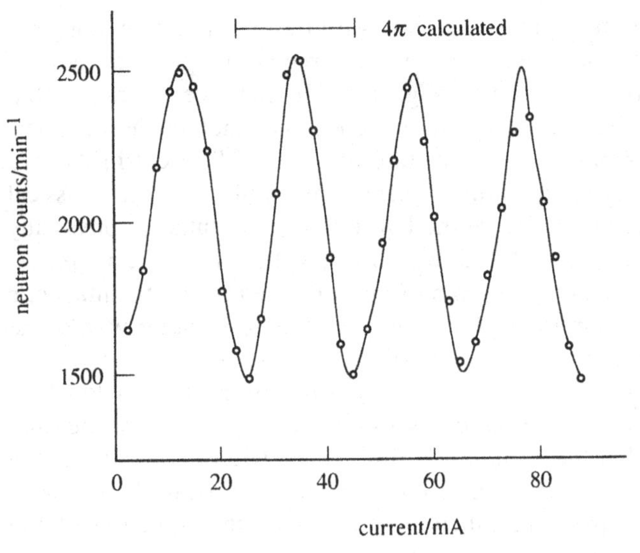

Fig. 10.10. *Neutron interferometer results of Rauch et al. (1975). The intensity of beam* III *is plotted against the current in the coil producing the magnetic field* **B**. *The results show that a 2π rotation of the neutron spin causes the wave function to change sign, and a 4π rotation is required to bring the wave function back to its original value.*

Comment

You may wonder what happens to the neutrons when the intensity, i.e. the number of neutrons, in beam III drops. The answer is that they go into beam IV – Fig. 10.8. To see how this comes about, we first note that, when a beam of neutrons undergoes partial Bragg reflection, the reflected beam is retarded in phase by $\pi/2$ with respect to the emerging non-reflected beam. (Bragg reflection arises from interference between scattered neutron waves, and the $\pi/2$ phase difference comes from summing all the scattered amplitudes in a phase–amplitude diagram.) If in Fig. 10.8 you follow the two beams that are combined to give beam III, you will see that each one is reflected twice – beam I at plates b and c, beam II at plates a and b. On the other hand, one of the components of beam IV undergoes one reflection, and the other one undergoes three reflections. Therefore, if the path lengths are equal, and there is no spin precession, the two components of III are in phase, while those of IV are out of phase. Any phase change now introduced affects the separate intensities of beams III and IV, but the sum of the two remains constant.

10.9 (a) The operator for resultant spin of the nucleus-neutron system is

$$\hat{\mathbf{T}} = \hat{\mathbf{J}} + \hat{\mathbf{S}}. \tag{1}$$

The three operators are vectors. Thus

$$\hat{T}^2 = \hat{J}^2 + \hat{S}^2 + 2\hat{\mathbf{J}}\cdot\hat{\mathbf{S}}. \tag{2}$$

If ψ is an eigenfunction of the operators $\hat{T}^2$, $\hat{J}^2$, $\hat{S}^2$, with respective eigenvalues $T(T+1)$, $J(J+1)$, $S(S+1)$, we have

$$\begin{aligned}\hat{\mathbf{J}}\cdot\hat{\mathbf{S}}\psi &= \tfrac{1}{2}(\hat{T}^2 - \hat{J}^2 - \hat{S}^2)\psi \\ &= \tfrac{1}{2}\{T(T+1) - J(J+1) - S(S+1)\}\psi.\end{aligned} \tag{3}$$

For the neutron $S = \frac{1}{2}$; the values of T are $J \pm \frac{1}{2}$. So the eigenvalues of $\hat{\mathbf{J}}\cdot\hat{\mathbf{S}}$ are

$$\tfrac{1}{2}\{T(T+1) - J(J+1) - S(S+1)\}$$

$$= \tfrac{1}{2}J \qquad T = J + \tfrac{1}{2}, \tag{4}$$

$$= -\tfrac{1}{2}(J+1) \quad T = J - \tfrac{1}{2}. \tag{5}$$

(b) If $|+\rangle$ is an eigenfunction of $\hat{b}$ with eigenvalue b^+, we have, using the result of (4),

$$\hat{b}|+\rangle = (A + B\hat{\mathbf{J}}\cdot\hat{\mathbf{S}})|+\rangle = (A + \tfrac{1}{2}JB)|+\rangle = b^+|+\rangle. \tag{6}$$

Thus

$$A + \tfrac{1}{2}JB = b^+. \tag{7}$$

Similarly

$$A - \tfrac{1}{2}(J+1)B = b^-, \tag{8}$$

whence

$$A = \{(J+1)b^+ + Jb^-\}/(2J+1),$$

$$B = 2(b^+ - b^-)/(2J+1). \tag{9}$$

(c) For the proton $J = \frac{1}{2}$. So

$$A = \tfrac{1}{4}(3b^+ + b^-), \quad B = b^+ - b^-. \tag{10}$$

When a neutron is scattered by a hydrogen molecule, each proton is the centre of a spherically symmetric scattered wave. If the wavelength of the neutrons is long compared to the distance between the two protons in the molecule, the two scattered waves are in phase, and the scattering length operator for the molecule is the sum of the scattering length operators for the two protons, i.e.

$$\hat{b}_{\text{mol}} = 2A + B\hat{\mathcal{J}}\cdot\hat{\mathbf{S}}, \tag{11}$$

where $\hat{\mathcal{J}}$ is the operator for the total spin of the protons in the molecule. Its eigenvalues are 1 for an ortho, and 0 for a para molecule. From (3), the eigenvalue of $\hat{\mathcal{J}}\cdot\hat{\mathbf{S}}$ is $\frac{1}{2}[T(T+1) - \mathcal{J}(\mathcal{J}+1) - S(S+1)]$, where T is now the total spin of the system of a neutron plus two protons.

For an ortho molecule, $\mathcal{J} = 1$, and $T = \frac{3}{2}$ or $\frac{1}{2}$. Thus the eigenvalues of $\hat{\mathcal{J}}\cdot\hat{\mathbf{S}}$ are

$$\begin{aligned} \tfrac{1}{2}\{T(T+1) - \mathcal{J}(\mathcal{J}+1) - S(S+1)\} &= \tfrac{1}{2} \quad & T = \tfrac{3}{2}, \\ &= -1 \quad & T = \tfrac{1}{2}. \end{aligned} \tag{12}$$

When the neutron is scattered by the ortho molecule, scattering occurs from all the states with the two T values. The degeneracy of each value is $2T+1$, i.e. 4 for $T = \frac{3}{2}$, and 2 for $T = \frac{1}{2}$. The eigenvalues of $\hat{\mathcal{J}}\cdot\hat{\mathbf{S}}$ have to be weighted by these degeneracies. Thus the average eigenvalue of $\hat{\mathcal{J}}\cdot\hat{\mathbf{S}}$ is

$$\langle\hat{\mathcal{J}}\cdot\hat{\mathbf{S}}\rangle = \{(4 \times \tfrac{1}{2}) - (2 \times 1)\}/6 = 0. \tag{13}$$

We shall also need the average eigenvalue of $(\hat{\mathcal{J}}\cdot\hat{\mathbf{S}})^2$, which is

$$\langle(\hat{\mathcal{J}}\cdot\hat{\mathbf{S}})^2\rangle = \{(4 \times \tfrac{1}{4}) + (2 \times 1)\}/6 = \tfrac{1}{2}. \tag{14}$$

We bring these results together. The measured differential scattering cross-section is the average eigenvalue of $(\hat{b}_{\text{mol}})^2$. For an ortho molecule, this is

$$\begin{aligned} \langle(\hat{b}_{\text{mol}})^2\rangle &= \langle(2A + B\hat{\mathcal{J}}\cdot\hat{\mathbf{S}})^2\rangle \\ &= 4A^2 + 4AB\langle\hat{\mathcal{J}}\cdot\hat{\mathbf{S}}\rangle + B^2\langle(\hat{\mathcal{J}}\cdot\hat{\mathbf{S}})^2\rangle \\ &= \tfrac{1}{4}(3b^+ + b^-)^2 + \tfrac{1}{2}(b^+ - b^-)^2. \end{aligned} \tag{15}$$

For the para molecule, $\mathcal{J} = 0$. So

$$\langle(\hat{b}_{\text{mol}})^2\rangle = 4A^2 = \tfrac{1}{4}(3b^+ + b^-)^2. \tag{16}$$

The total scattering cross-section σ_{tot} is obtained by multiplying the differential scattering cross-section by 4π. We also insert the fact that the bound scattering lengths b^+ and b^-, refer to a fixed proton. For protons in a hydrogen molecule the scattering length is $\frac{2}{3}$ times the bound scattering length. Therefore

$$\sigma_{\text{tot}} = \frac{16\pi}{9}\langle(\hat{b}_{\text{mol}})^2\rangle, \tag{17}$$

which, together with (15) and (16), gives the required results.

Comments

(1) The result in (13), that the average eigenvalue of $\hat{\mathcal{J}}\cdot\hat{\mathbf{S}}$ is zero, reflects the fact that there is no correlation between the spin of the neutron and the spins of the protons. Only if the spins of both the neutrons and the protons were aligned could there be such a correlation. In which case the weighting factors would be different from those in (13), and $\langle\hat{\mathcal{J}}\cdot\hat{\mathbf{S}}\rangle$ would not, in general, be zero.

(2) The scattering lengths, b^+ and b^-, relate to a bound nucleus, as in a solid, and are sometimes known as *bound* scattering lengths. If the nucleus is not bound, the scattering must be treated in the centre-of-mass system. The result is the same as if the nucleus were fixed, but the mass m_{n} of the neutron must be replaced by the reduced mass μ of the neutron–molecule system. The effect of this on the scattering length can be seen from the results of Problem 9.8, where it is shown that the potential $V(\mathbf{r})$ for the scattering of neutrons by a nucleus can be expressed as

$$V(\mathbf{r}) = \frac{2\pi\hbar^2 b}{m_{\text{n}}}\delta(\mathbf{r}). \tag{18}$$

The potential is the same whether the nucleus is fixed or not. So we have the result that the scattering length b_{mol} for the proton in the hydrogen molecule is related to the bound scattering length b, by

$$\frac{b_{\text{mol}}}{\mu} = \frac{b}{m_{\text{n}}}. \tag{19}$$

The mass m_{mol} of the molecule is approximately $2m_{\text{n}}$. So the relation

$$\frac{1}{\mu} = \frac{1}{m_{\text{mol}}} + \frac{1}{m_{\text{n}}} = \frac{3}{2m_{\text{n}}} \tag{20}$$

gives

$$b_{\text{mol}} = \tfrac{2}{3}b. \tag{21}$$

10.10 (a) The Hamiltonian is

$$\hat{H} = \frac{1}{2m}\hat{p}^2 = -\frac{\hbar^2}{2m}\frac{\mathrm{d}^2}{\mathrm{d}x^2}. \tag{1}$$

So the Schrödinger equation $\hat{H}\psi = E_0\psi$ is

$$\frac{\mathrm{d}^2\psi}{\mathrm{d}x^2} + \mathrm{k}^2\psi = 0, \tag{2}$$

where

$$k^2 = \frac{2mE_0}{\hbar^2}. \tag{3}$$

The solution is

$$\psi = c\exp(\pm\mathrm{i}kx), \tag{4}$$

where c is a constant. Since x and $x + 2\pi a$ are the same point, ψ must be the same at the two values of x. Therefore

$$k = \frac{n}{a}, \tag{5}$$

where n is zero or a positive integer. The constant c is fixed by the normalisation condition

$$\int_0^{2\pi a} |\psi|^2\,\mathrm{d}x = 2\pi a c^2 = 1. \tag{6}$$

From (3) and (5)

$$E_0 = \frac{n^2\hbar^2}{2ma^2}. \tag{7}$$

(b) The magnetic dipole moment is

$$\mu = (\text{area of orbit}) \times \text{current} = \pm\frac{\pi a^2 e}{\tau}, \tag{8}$$

where τ is the period of revolution. If v is the velocity of the particle, then

$$\frac{1}{\tau} = \frac{v}{2\pi a} = \frac{p}{2\pi a m} = \frac{\hbar k}{2\pi a m} = \frac{n\hbar}{2\pi a^2 m}. \tag{9}$$

Thus

$$\mu = \pm \frac{ne\hbar}{2m}. \tag{10}$$

(c) Stokes' theorem, together with the relation $\mathbf{B} = \text{curl}\,\mathbf{A}$, gives

$$\oint \mathbf{A}\cdot \mathrm{d}\mathbf{l} = \int \mathbf{B}\cdot \mathrm{d}\mathbf{S}, \tag{11}$$

where $\mathrm{d}\mathbf{l}$ is an element of length along the orbit, and $\mathrm{d}\mathbf{S}$ is an element of area in the surface enclosed by the orbit – Fig. 10.11(a). Since $\mathbf{A}$ is tangential to the orbit and of constant magnitude round it, the value of the line integral is $2\pi a A$. The field $\mathbf{B}$ is perpendicular to the surface of the orbit and is uniform. So the value of the surface integral is $\pi a^2 B$. Equating the two values gives

$$A = \frac{aB}{2}. \tag{12}$$

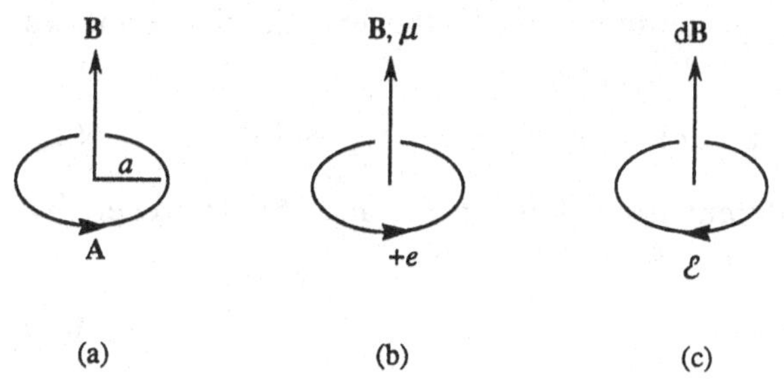

Fig. 10.11. *(a) Relation between magnetic field* **B** *and vector potential* **A**. *(b) Motion of charge* $+e$ *for the state* ψ_+*, and the magnetic dipole moment* $\boldsymbol{\mu}$ *resulting from the motion. (c) An increase* $\mathrm{d}\mathbf{B}$ *in the magnetic field in time* $\mathrm{d}t$ *produces an electric field* $\mathcal{E}$ *in the direction shown.*

(d) The Hamiltonian is

$$\hat{H} = \frac{1}{2m}(\hat{p} - eA)^2 = \frac{1}{2m}\left(\frac{\hbar}{\mathrm{i}}\frac{\mathrm{d}}{\mathrm{d}x} - eA\right)^2. \tag{13}$$

The Schrödinger equation is

$$-\frac{\hbar^2}{2m}\frac{\mathrm{d}^2\psi}{\mathrm{d}x^2} + \frac{\mathrm{i}e\hbar A}{m}\frac{\mathrm{d}\psi}{\mathrm{d}x} + \frac{e^2A^2}{2m}\psi = E\psi. \tag{14}$$

The function

$$\psi = \exp(qx) \tag{15}$$

is a solution of (14), if q satisfies

$$-\frac{\hbar^2 q^2}{2m} + \frac{ie\hbar Aq}{m} + \frac{e^2 A^2}{2m} - E = 0, \tag{16}$$

i.e.

$$q^2 + 2uq + w = 0, \tag{17}$$

where

$$u = -\frac{ieA}{\hbar} = -ij, \tag{18}$$

$$w = \frac{2mE}{\hbar^2} - \frac{e^2 A^2}{\hbar^2} = k^2 - j^2, \tag{19}$$

$$k^2 = \frac{2mE}{\hbar^2}. \tag{20}$$

From (17) to (19)

$$q = -u \pm \sqrt{(u^2 - w)} = i(j \pm k). \tag{21}$$

Ignoring the normalising constant outside the exponential term, we have two solutions

$$\psi_+(x) = \exp\{i(j + k)x\}, \quad \psi_-(x) = \exp\{i(j - k)x\}. \tag{22}$$

These functions look different from those in (4) and (5). However, we have the condition that

$$\psi_\pm(x) = \psi_\pm(x + 2\pi a), \tag{23}$$

which leads to

$$j + k = \frac{n}{a} \quad \text{for } \psi_+, \quad j - k = -\frac{n}{a} \quad \text{for } \psi_-. \tag{24}$$

Inserting these values in (22) gives the same functions as (4) with (5). However, the values of k, and hence, from (20), those of the energy, are different.

For ψ_+

$$k_+ = \frac{n}{a} - j, \quad E_+ = \frac{\hbar^2}{2m}\left(\frac{n}{a} - j\right)^2. \tag{25}$$

For ψ_-

$$k_- = \frac{n}{a} + j, \quad E_- = \frac{\hbar^2}{2m}\left(\frac{n}{a} + j\right)^2. \tag{26}$$

Thus

$$E_\pm = E_0 + E_1 + E_2, \tag{27}$$

where

$$E_1 = \mp \frac{n\hbar^2 j}{ma}, \quad E_2 = \frac{\hbar^2 j^2}{2m}. \tag{28}$$

(e) From the relations $\hbar j = eA$, and $A = aB/2$, we have, for the state ψ_+,

$$E_1 = -\frac{n\hbar^2 j}{ma} = -\frac{ne\hbar}{2m} B = -\mu B \tag{29}$$

from (10). Thus E_1 corresponds to the energy of interaction between the previously existing magnetic dipole and the applied field **B**, i.e. to the paramagnetic energy. The directions of $\boldsymbol{\mu}$ and **B** are the same – Fig. 10.11(b) – so $E_1 = -\boldsymbol{\mu}\cdot\mathbf{B}$ is a negative quantity. For the state ψ_-, the particle rotates in the opposite sense, $\boldsymbol{\mu}$ is in the opposite direction, and E_1 is positive.

The expression for E_2 is

$$E_2 = \frac{\hbar^2 j^2}{2m} = \frac{e^2 A^2}{2m} = \frac{e^2 a^2 B^2}{8m}. \tag{30}$$

The diamagnetic effect is the interaction of the change in the magnetic dipole moment produced by the applied magnetic field with the magnetic field itself. We calculate the interaction energy classically by considering the field gradually increasing from zero to its final value, which for the moment we denote by B_0. At some intermediate time an increase $\mathrm{d}B$ in time $\mathrm{d}t$ produces an electric field $\mathscr{E}$ acting round the orbit, whose magnitude is given by Faraday's law of induction, i.e.

$$\pi a^2 \frac{\mathrm{d}B}{\mathrm{d}t} = 2\pi a \mathscr{E}. \tag{31}$$

The direction of $\mathscr{E}$ is shown in Fig. 10.11(c). If the particle is in state ψ_+, and has velocity v, the electric field causes a change in the momentum

$$m\,\mathrm{d}v = -\mathscr{E} e\,\mathrm{d}t = -\frac{ea}{2}\mathrm{d}B. \tag{32}$$

From (8) and (9) the magnetic dipole moment of the particle is

$$\mu = \frac{\pi a^2 e}{\tau} = \frac{eav}{2}. \tag{33}$$

So the change in momentum in (32) gives a change in μ of

$$\mathrm{d}\mu = \frac{ea}{2}\mathrm{d}v = -\frac{e^2 a^2}{4m}\mathrm{d}B. \tag{34}$$

The total energy change as B is increased from 0 to B_0 is

$$E_{\text{dia}} = -\int_0^{B_0} B\,\mathrm{d}\mu = \frac{e^2a^2}{4m}\int_0^{B_0} B\,\mathrm{d}B = \frac{e^2a^2B_0^2}{8m}. \tag{35}$$

Dropping the subscript in B_0 we see that $E_2 = E_{\text{dia}}$. The expression for E_2 does not contain the Planck constant, and again the quantum and classical results agree.

Comments

(1) This problem illustrates the modification of the relation between wavelength and momentum caused by a magnetic field. The basic relation (in one-dimension) is

$$p = mv + eA. \tag{36}$$

p is known as the *canonical momentum*, and its operator is $(\hbar/\mathrm{i})\,(\mathrm{d}/\mathrm{d}x)$. The quantity mv is the ordinary or *mechanical momentum*, and its operator is $(\hbar/\mathrm{i})\,(\mathrm{d}/\mathrm{d}x) - eA$.

The de Broglie relation is

$$\lambda = \frac{\hbar}{p}. \tag{37}$$

p does not depend on A. So the application of a magnetic field does not change the wavelength of a particle. On the other hand, the field does change the velocity of the particle, and hence its kinetic energy. The Hamiltonian depends on A according to (13). This is for a simple one-dimensional problem with constant A. In a three-dimension problem, p becomes the three-dimensional operator

$$\mathbf{p} = \frac{\hbar}{\mathrm{i}}\left(\mathbf{e}_x\frac{\partial}{\partial x} + \mathbf{e}_y\frac{\partial}{\partial y} + \mathbf{e}_z\frac{\partial}{\partial z}\right), \tag{38}$$

where $\mathbf{e}_x$, $\mathbf{e}_y$, $\mathbf{e}_z$ are unit vectors in the x, y, z directions, and $\mathbf{A}$ is, in general, a function of position.

(2) Note that the sign of the diamagnetic energy E_2 is positive both for ψ_+ and for ψ_-. We can see this from the physical picture. For ψ_+, $\boldsymbol{\mu}$ and $\mathbf{B}$ are in the same direction – Fig. 10.11(b). The magnitude of $\boldsymbol{\mu}$ decreases as $\mathbf{B}$ is applied, so the potential energy of the dipole in the presence of $\mathbf{B}$, which is negative, becomes less negative, i.e. it increases. For ψ_-, $\boldsymbol{\mu}$ and $\mathbf{B}$ are in opposite directions. The electric field $\mathcal{E}$ acts in the same direction as before, but it now accelerates the particle and increases the magnitude of $\boldsymbol{\mu}$. Therefore the potential energy of the dipole, which is positive, increases.

The diamagnetic energy is also positive for a negatively charged particle. This follows from the fact that e enters the expression for E_2 as e^2. You can reach the same conclusion more lengthily by following through the reasoning of the previous paragraph for a negative charge.

References

Bloch, F. 1932. *Zeitschrift für Physik*, **74**, 295.

Bransden, B. H. and Joachain, C. J. 1983. *Physics of Atoms and Molecules*, Longman.

Cohen, E. R. and Taylor, B. N. 1986. *The 1986 Adjustment of the Fundamental Physical Constants, Codata Bulletin Number 63*, Pergamon Press.

Condon, E. U. and Shortley, G. H. 1963. *The Theory of Atomic Spectra*, Cambridge University Press.

Davydov, A. S. 1965. *Quantum Mechanics*, Pergamon Press.

Geiger, H. and Marsden, E. 1913. *Philosophical Magazine*, **25**, 604.

Glauber, R. J. 1963. *Physical Review*, **131**, 2766.

Goldstein, H. 1980. *Classical Mechanics*, 2nd ed., Addison-Wesley.

Haken, H. and Wolf, H. C. 1987. *Atomic and Quantum Physics*, 2nd ed., Springer-Verlag.

Kittel, C. 1986. *Introduction to Solid State Physics*, 6th ed., John Wiley & Sons.

Loudon, R. 1983. *The Quantum Theory of Light*, 2nd ed., Oxford: Clarendon Press.

Mathews, J. and Walker, R. L. 1970. *Mathematical Methods of Physics*, 2nd ed., Benjamin.

Mott, N. F. and Massey, H. S. W. 1965. *The Theory of Atomic Collisions*, 3rd ed., Oxford: Clarendon Press.

Rauch, H., Zeilinger, A., Badurek, G., Wilfing, A., Bauspiess, W. and Bonse, U. 1975. *Physics Letters*, **54A**, 425.

Schiff, L, I. 1968. *Quantum Mechanics*, 3rd ed., McGraw-Hill.

Schubert, M. and Wilhelmi, B. 1986. *Nonlinear Optics and Quantum Electronics*, John Wiley & Sons.

Summhammer, J., Badurek, G., Rauch, H., Kischko, U. and Zeilinger, A. 1983. *Physical Review*, **A27**, 2523.

Index